# Foreign Policy in Greenland

*Foreign Policy in Greenland* will shed light on what kind of competences Greenland has in IR and how Greenland undertakes foreign policy, while still being a self-governing territory underneath Danish jurisdiction. Even though Denmark has the ultimate power when it comes to foreign- and security policy, Greenland has received more autonomy in this field in recent years. Back in 1995 a permanent Danish/Greenlandic official group was established to discuss all foreign- and security matters that were related to Greenland. In 2003, Greenland and Denmark signed a joint declaration in Itilleq regarding the Greenlandic involvement in international affairs. Since then, the Government of Greenland has written a foreign policy report every year to outline the activities of Greenlandic international relations.

This book will undertake a systematic analysis of the foreign policy papers since 2003. What kind of activities have Greenland undertaken in international affairs? What are the objectives? Can Greenland be an influential player in international relations? How far is the self-government stretched or limited in relation to international relations?

This book will be of interest for those with an interest in international affairs and law as well as those with an interest in international politics and foreign policy.

**Maria Ackrén** is currently Professor of Political Science with a special focus on island autonomy at Ilisimatusarfik/University of Greenland since January 2024.

## Routledge Research in Polar Regions
## Series Editor: Timothy Heleniak
*Nordregio International Research Centre, Sweden*

The Routledge series in Polar Regions seeks to include research and policy debates about trends and events taking place in two important world regions: the Arctic and Antarctic. Previously neglected periphery regions, with climate change, resource development and shifting geopolitics, these regions are becoming increasingly crucial to happenings outside these regions. At the same time, the economies, societies and natural environments of the Arctic are undergoing rapid change. This series seeks to draw upon fieldwork, satellite observations, archival studies and other research methods which inform about crucial developments in the Polar regions. It is interdisciplinary, drawing on the work from the social sciences and humanities, bringing together cutting-edge research in the Polar regions with the policy implications.

**Indigenous Peoples, Natural Resources and Governance**
Agencies and Interactions
*Edited by Monica Tennberg, Else Grete Broderstad, and Hans-Kristian Hernes*

**The Future of the Arctic Human Population: Migration in the North**
*Edited by Nafisa Yeasmin, Satu Uusiautti, Timo Koivurova, and Timothy Heleniak*

**Distance, Equity and Older People's Experiences in the Nordic Periphery**
Centering the Local
*Edited by Shahnaj Begum, Joan R. Harbison, Päivi Naskali, and Minna Zechner*

**The Significance of Sámi Rights**
Law, Justice, and Sustainability for the Indigenous Sámi in the Nordic Countries
*Edited by Dorothée Cambou and Øyvind Ravna*

**Human Security through the New Traditional Economy in the Arctic**
*Edited by Gorm Winther and Ívar Jónsson*

For more information about this series, please visit: www.routledge.com/Routledge-Research-in-Polar-Regions/book-series/RPR

# Foreign Policy in Greenland

## 20 Years of Development

**Maria Ackrén**

Routledge
Taylor & Francis Group
LONDON AND NEW YORK

First published 2025
by Routledge
4 Park Square, Milton Park, Abingdon, Oxon OX14 4RN

and by Routledge
605 Third Avenue, New York, NY 10158

*Routledge is an imprint of the Taylor & Francis Group, an informa business*



*British Library Cataloguing-in-Publication Data*
A catalogue record for this book is available from the British Library

ISBN: 9781032481500 (hbk)
ISBN: 9781032481524 (pbk)
ISBN: 9781003387657 (ebk)

DOI: 10.4324/9781003387657

Typeset in Times New Roman
by codeMantra

# Contents

# Figures

# Tables

# About the Author

**Maria Ackrén** is currently Professor of Political Science with a special focus on island autonomy at Ilisimatusarfik/University of Greenland since January 2024. Maria is originally from the Åland Islands in Finland, but has lived and worked in Nuuk, Greenland since 2011. In 2009, in her PhD dissertation, she was dealing with a comparative study of all territorial autonomies in the world. The dissertation was a monograph. Since then, Maria has moved into the field of international relations and has been concentrating on Arctic relations and especially Greenland. She has published various publications in both high-quality journals and anthologies and more specialized journals and books related to Arctic and island studies. Most of the works are related to self-governance issues, international relations, regional parties and elections, the management of natural resources in Greenland, and comparative methods. Maria is also elected as IASSA president for 2024–2027 and is Head of Research for Nasiffik – Center for Foreign- and Security Policy at Ilisimatusarfik/University of Greenland.

# Acknowledgments

The author would like to thank Greenland Research Council for some seed money for the project. The funding was crucial to start the book project overall. The author would also like to thank Head of Institute Annemette Nyborg Lauritsen for accepting a semester of sabbatical during the writing process. Furthermore, the author would like to thank all colleagues at the Department of Arctic Social Sciences and Economics for their comments and suggestions of ideas that have been discussed in various formats. The author would also like to thank the participants on the panel session of Arctic security and international relations at Polar Law Symposium on 25th of September 2024 in Östersund, Sweden who posed questions during the presentation of some elements of this book. I am also deeply thankful to the editors at Routledge and reviewers on an earlier draft of this book. Without all the constructive comments this book would not have reached the same quality and conclusions.

Nuuk, December 2024

# 1 Introduction

Greenland is said to be the largest island in the world with a total area of approximately 2.2 $km^2$, but with a quite small population of 56,000. About 80% of the country is covered by ice or glaciers. Greenland is strategically located on the North American continent between the Arctic Ocean and the North Atlantic Ocean, northeast of Canada. The northernmost point of Greenland, Oodaap Qeqertaa is lying 706 km from the North Pole. The southernmost point is Cape Farewell which is about the same latitude as Oslo in Norway. The population lives in towns and settlements along the coast but close to one third of the population lives in the capital, Nuuk (19,880 in 2024) (*Greenland in Figures 2024*).

Greenland was connected to the outside world already in 1261, when the Norse Greenlanders in southwest Greenland agreed to be part of the Norwegian Kingdom. However, this was not a formal agreement and the establishment of the Kalmar Union in 1397 and especially the subsequent union between Denmark and Norway in 1536 established a real connection between Greenland and the Danish Kingdom, however this was not in a physical form (Høegh, 2024:182–183). It took until 1721 before Greenland was reconnected with the Kingdom of Denmark-Norway with the arrival of Hans Egede a Danish-Norwegian missionary.

Greenland was first ruled as a colony 1721–1953, then became integrated as a county within the Danish jurisdiction 1953–1978. In 1978 a referendum took place where the alternative was to become a self-governing territory in the same way as the Faroe Islands. Most of the Greenlanders wanted the autonomy alternative. During the 2000s a revision of the Home Rule was made and in 2009 a new Self-Government Act gave Greenland more competencies and room for maneuver especially within international relations. The new Act also includes an acknowledgement that the Greenlandic people is defined as a people according to international law and has therefore the opportunity to opt out of the Danish Kingdom if the people wish to become independent.

Politically, Greenland has its own legislative assembly, Inatsisartut with 31 members, which are elected every fourth year. The Government of Greenland, Naalakkersuisut is the executive branch and there are also some Greenlandic courts at the first and second instance. Greenland has a wide range of competencies and law-making capacity in several political areas, including some issues of international affairs. Greenland can sign bilateral treaties with foreign powers and be part

DOI: 10.4324/9781003387657-1

of international organizations, where not just states are members. Furthermore, Greenland has its own "mini embassies" abroad in form of foreign offices of representation, where Greenlandic diplomats work. It is important to stress that the Greenland's own diplomats have a deep insight into the Greenlandic society and its political reality. They have all been growing up with the Greenlandic language and culture. The representations abroad take their instructions from the Greenland Ministry of Statehood and Foreign Affairs in Nuuk, and work in close collaboration with the Danish diplomats at the embassies and the Ministry of Foreign Affairs in Copenhagen if issues fall outside Greenland's jurisdiction (Høegh, 2024:188).

Greenland has become a hot spot in the Arctic due to climate change and the assets of natural resources that can be found both on land and in the sea around Greenland. Some of these assets are of great interests for larger powers, such as, China, the USA, and others who have an interest in especially dual-use minerals (rare earth elements, uranium, and the like). Large-scale mining projects and oil and gas development have been in focus in the recent decade; however, no large-scale project has yet been established. There are some smaller mining projects at hand and the oil and gas adventure was on its hype during the 2010s, but most international companies have withdrawn their activities in the Arctic. However, the Government of Greenland is actively trying to find ways of alternating its economy, since the fisheries are the largest business to date and the fluctuations in this business are challenging for a small economy. The tourism sector is also an opportunity for future endeavors, while new airport strips are extended and built in Nuuk, Ilulissat and Qaqortoq. The International Airport in Nuuk has recently opened as of 28th of November 2024 with its first direct flight between Copenhagen and Nuuk. During the summer of 2025, we will experience several other operators in form of SAS, Icelandair, United Airlines and Canadian North, which will open direct routes from Copenhagen, Newark (USA) and Iqaluit (Canada). This will enhance tourism opportunities and Greenland become much closer to the world at large.

## 1.1 The Arctic Context

The Arctic occupies about one-sixth of the globe and is home to approximately four million people. There has been a perception that many of the security concerns that occupy most of the world are not widely found in the Far North. The climate and geography of the Arctic have been seen as too hostile for large-scale state-centric security activity, except in rare cases. Since the end of the Cold War there has been an effort among the Arctic states to create a space described as "high north, low tension" (Hoogensen Gjørv, 2020:1–2). However, the Arctic has been a scene for various interests over the centuries. In the past interests were focused on research expeditions and hunting expeditions, and during World War II it was a strategic arena placed between allied powers and the enemy.

At the end of World War II, air power was of vital importance on a global scale. Given the increased speed and range of American and Russian bombers, the United States and the Soviet Union had overlapping security areas across the Arctic. The areas where possible collisions could occur were above Alaska and Norway. It became necessary to establish a form of defense mechanism in the North, which would cover not only the United States, but Canada too. This led to the creation of the North American Aerospace Defense Command (NORAD) with headquarters at Ent Air Force Base in Colorado and a command center under Cheyenne Mountain in the same state in 1957. The continent became covered for surveillance purposes by Distant Early Warning (DEW) Line radar stations stretched from Alaska to Greenland. These were backed up by the Mid-Canada and Pine Tree Lines (Henriksen, 2020:15). The purpose of the DEW Line was to monitor the Arctic airspace and to be able to deter and provide advance notice of potential attacks (Padrtova, 2020:35). Today, the North Warning System (NWS), a joint US-Canadian undertaking, is based on these antecedents. The North Atlantic Treaty Organization (NATO) established in 1949 also became a vital forum, where Denmark, Iceland and Norway were included as members also covering a large part of the Arctic region together with USA and Canada.

A renewed interest in the Arctic region started in the 1980s amongst the superpowers, the Soviet Union, and the USA. The 1980s became again a strategic period where the two superpowers enhanced their military activity in the region (Young, 1985:160). Another feature leading to more interest was the rapid industrialization in the Far North. Exploration and exploitation of oil, gas and minerals became of vital importance for both superpowers' security (Archer, 1988:139; Young, 1985:160). However, it was not only the superpowers that upgraded their military preparedness in the northern regions. Denmark/Greenland, Iceland, and the Faroe Islands also improved their defense installations, as well as Norway and Canada (Archer, 1988:138). Traditionally, the USA has not had a focus on the Arctic within the foreign- and security policy since more attention has been directed towards the Middle East and other areas. Russia on the other hand possesses crucial infrastructure and relevant operational capacities in the Arctic (Wegge, 2011). In recent years, the USA has enhanced its interest towards the Arctic. This has been seen in the relationship towards Greenland, where Pituffik/Thule Air Base[1] has been upgraded and modernized and the renewed activities in Iceland, where the Keflavik base was closed back in 2006, but a renewed interest of opening the base has been realized.

Another event that had a major impact on the development in the Arctic was the so-called Murmansk Initiative, beginning with the famous speech by Mr. Gorbachev in Murmansk 1 October 1987. In this speech, he mentioned six areas for international discussion amongst the states and regions with interests in the Arctic. The six areas of interest were:

> a Nordic nuclear-free zone in Northern Europe; limited naval activity in the seas adjacent to Northern Europe; peaceful cooperation in exploiting the resources of the North and the Arctic; the scientific study of the Arctic; cooperation among Northern countries in environmental protection; and the opening up of the Northern Sea Route to foreign vessels.
>
> (Archer, 1988:140)

In the West, the "zone of peace" concept was in line with the capitalist strategy of desecuritizing economic interaction and to spread democracy, while for the East the concept was frequently used synonymously as demilitarization, but also as a concept for bilateral and multilateral cooperation within soft power areas (or non-military areas) (Åtland, 2008). The Murmansk speech was meant as a trigger for cooperation among Arctic states in several areas, such as, resource extraction, scientific exploration, indigenous peoples, environmental protection, and marine transportation (Åtland, 2008:300). One outcome of the Murmansk Initiative was the establishment of the International Arctic Science Committee (IASC), which came into being after a meeting in Stockholm 1988 where all the eight Arctic states[2] were participating. This body was to be a focal point for discussions, cooperation, and interaction among Arctic scientists. The matters of interest surrounded on issues related to Arctic lands, seas, atmosphere, and space (Friedheim, 1988:507).

A third issue related to Arctic affairs was enhanced cooperation among the Arctic states because of the end of the Cold War in the late 1980s and early 1990s. In 1989, Finland invited officials from the eight Arctic states to Rovaniemi in the Finnish Lapland to discuss cooperative measures to protect the Arctic environment. The efforts that followed set up the Arctic Environmental Protection Strategy (AEPS). The AEPS was establishing four working groups: the Arctic Monitoring and Assessment Program (AMAP); the Conservation of Arctic Flora and Fauna (CAFF); the Emergency Prevention, Preparedness and Response (EPPR); and the Protection of the Arctic Marine Environment (PAME) (Pedersen, 2012). This was the first initial step towards a more institutionalized cooperation that later developed into the establishment of the Arctic Council. The Arctic Council became established in 1996 and fused the AEPS programs and working groups into its structure with a supplementary initiation of a Sustainable Developing Program (Young, 2002:289). The Arctic Council has become seen as the most prominent forum for Arctic issues, since it also includes the indigenous Permanent Participants consisting of six indigenous organizations representing the various indigenous groups in the Arctic. The input from indigenous and regional peoples are of utmost importance within the circumpolar cooperation (Heininen and Nicol, 2007). Greenland is participating within the Danish delegation on an equal footing as Denmark and the Faroe Islands. Furthermore, Greenland is also taken part in the Inuit Circumpolar Council (ICC) and is therefore also represented as a Permanent Participant. Supplementary to this the Ministry of Foreign Affairs in Greenland takes an active role within several of the working groups within the Arctic Council (Ackrén, 2020).[3]

The starting point of a more hyped interest towards the High North in the 2000s started with the much-published event of the Russian flag planting on the bottom of the central Arctic basin in summer 2007. This started new narratives of 'Scramble for the Arctic' (as earlier 'Scramble of Africa' during the nineteenth century) and 'Exploitation of the last frontier' and similar connotations (Powell and Dodds, 2014:4). At the same time, a lot of activities were centered around research and varying projects linked to the International Polar Year (2007–2009), and the US Geological Survey released a report stating that there might be 90 billion barrels of undiscovered oil north of the Arctic Circle (Powell and Young, 2014:4; US Geological Survey, 2008). This gave fuel to headlines, such as, "Rush for Arctic's resources provoke territorial tussles" and "Arctic black gold mapped" in international media outlets (see for example, the Guardian, 2011 and Nature, 2009).

Foreign and security policy has become a topical issue in an Arctic context during recent decades. The attention towards the Arctic and Greenland takes many forms and relates to issues, such as, climate change, natural resources, new shipping lanes, indigenous affairs, geopolitics, and postcolonial contexts. In academia, we see an enhanced literature on the various topics related to Greenland's international role (Ackrén and Nielsen, 2021). Authors have been looking at the identity of Greenland within international relations (Jacobsen and Gad, 2018), the relationships between Denmark and Greenland (Rahbek-Clemmensen and Sørensen, 2021) and how the growing foreign policy is outlined in a Greenlandic context (Kleist, 2019). In a recent anthology published in 2024, the securitization theory has been used of the so-called Copenhagen School to cover Arctic and Greenland's security (see Jacobsen et al., 2024). Furthermore, recent articles have been concentrated around Greenland-US relationships with theories of action space and two-level games (Olsvig, 2022a; Olsvig, 2022b). This book will be limited to focus on Greenland and its role related to international relations (IR). Greenland is as mentioned previously a self-governing territory belonging to Denmark with a wide range of competencies, when it comes to internal matters, but in recent years Greenland has also enhanced its role internationally. The interest is to shed light on the role of subnational jurisdictions, such as Greenland plays in IR. How have the competencies evolved over time? How is Greenland making its voices heard in IR? What constraints might occur between Denmark and Greenland within foreign- and security policy?

## 1.2 The Greenlandic Case

Greenland's formal foreign policy competencies are regulated in the Danish constitution and the Self-Government Act from 2009. The Self-Government Act stipulates that the Government of Greenland can negotiate and sign bilateral treaties with foreign powers regarding issues that are within the competencies of Greenland. The Government of Greenland should notify Denmark about the negotiations and whether a treaty will be signed or not. This notification procedure is also valid the other way, where Denmark might conclude international agreements that might have crucial importance for Greenland (Kleist, 2019). Greenland has four foreign representation offices abroad and an office of representation in Copenhagen as well. The representation office in Brussels takes care of EU matters, the representation office in Washington D.C. covers interests between Greenland and the USA and Canada and the representation office in Reykjavik takes care of interests between Greenland and Iceland. The representation office in Copenhagen covers the relationships between Greenland and Denmark and has some tasks related to the diplomatic corps in Denmark. The Self-Government Act is regulating the issue of establishing foreign representations (Kleist, 2019). A new representation office has recently opened in 2021 in Beijing, China to cover the interests between Asian countries and Greenland. Most of the work the representation offices do is to build up diplomatic relations and develop networks and contacts for business and trade.

Denmark is still the overall power when it comes to foreign- and security policy. This means that Greenland cannot become a member in international organizations where only states are members. However, issues emerging in climate change, claims for extended national jurisdictions regarding the continental shelf, exploration and exploitation of natural resources, and the protection of the indigenous Inuit population are all matters that have demanded and continuous to demand complex negotiations between Danish and Greenlandic authorities (Ackrén and Jakobsen, 2015).

At the national level there are national sessions taken place between the Danish Prime Minister and the Greenlandic Premier (as well as the Faroese Premier) annually. These sessions are held in all three parts of the Kingdom of Denmark. There are also meetings at the official level between officials from different departments[4] within the governments. The Greenlandic government has a foreign office, which handles the coordination and contact between Nuuk and the foreign Greenlandic representations abroad. The foreign office is functioning as an advisory board in international relations (Ackrén, 2014). The foreign office has a broad range of issues on its agenda, such as, Arctic cooperation, the EU, the UN, Nordic cooperation, Indigenous Peoples issues, the Joint Committee between the USA and Denmark/Greenland, foreign- and security policy, and foreign commercial matters. In the parliament (*Inatsisartut*) there is also a foreign- and security policy committee, which was established in 1988. This committee is functioning as a mirror agency to the foreign office with parliamentarians as members (Ackrén, 2014). Greenland has for decades worked towards an enhanced role within international relations to give its own voice within those matters especially important for Greenland, such as, fishery, hunting, indigenous issues and so forth (Jacobsen and Gad, 2018). This has evolved into Greenland's own strategy for foreign-, security and defense policy published in February 2024 (see *Greenland's Foreign, Security and Defense Policy 2024–2033*).

This book will give a detailed description of the Greenlandic international relations related to foreign- and security policy. Domestic and international politics are intertwined in Greenland as domestic politics are affected by external influences and shape the politics daily. At the same time Greenland's paradiplomatic behavior affects Denmark's bilateral relations with other nations and the general politics of the Arctic. Greenland's endeavors of becoming an independent state also influences the relationships towards Denmark and other countries. As Jacobsen, Wæver and Gad (2024) is stating, Greenland is "one of the most difficult polities to fit in traditional categories of IR" (Jacobsen et al., 2024:1). In this sense Greenland is an interesting case to study and might be a forerunner for other subnational jurisdictions within IR.

## 1.3 Research Area and Problem

The book will outline Greenland's role in international relations by specifically looking at the foreign policy papers that the Government of Greenland is outlining annually. The timeline will be concentrated on the period from 2003 to 2023, a

20-year period, to be able to see the effects of how the foreign- and security policy area has evolved from Home Rule to Self-Government and beyond. The major themes in the book will be the discussion around foreign policy analysis, paradiplomacy, shelter theory and two-level games theory as a theoretical framework for shedding light on how Greenland navigate within IR as still being part of the Kingdom of Denmark. The methodology will be document and content analysis, as well as case study research, since this will give the reader an in-depth analysis of Greenland's international relations. This form of analysis has not been undertaken earlier. The book will therefore shed new lights on Greenland's role within IR.

The book will undertake a systematic analysis of the foreign policy papers from 2003 and onwards. What kind of activities are Greenland undertaken in international affairs? What are the objectives? Can Greenland be an influential player within IR? How far is the self-government stretched or limited in relation to IR?

## 1.4 Research Design

This book will take on a detailed case study of Greenland in relation to IR and foreign policy. Document and content analysis and case study research will be the ultimate methodology used in this respect. The main documents are the earlier mentioned annual reports coming from the Government of Greenland and other relevant documents. Other material will consist of academic literature, media outlets, and relevant reports. The methodology will be further described in Chapter 3.

The connection between the theoretical framework and the methodology part to answer the research questions are related to the three levels of analysis of individual, state/nation, and systematic level. It is possible to illustrate the research design as follows (Figure 1.1):

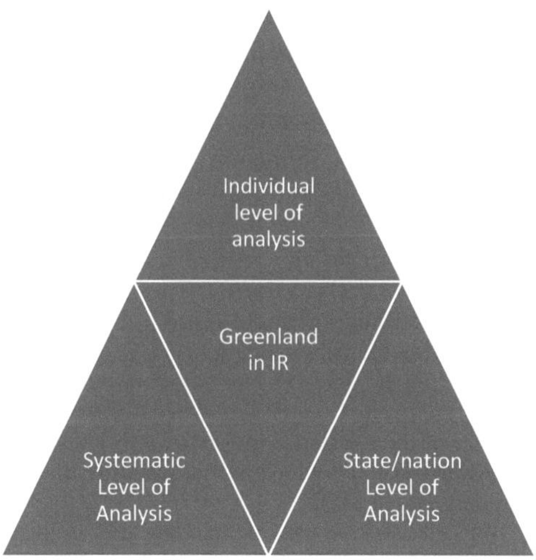


*Figure 1.1* Research Design.

The individual level of analysis is related to answering the question of what the objectives are within IR from officials and politicians in Greenland. This will be related to the overall theory of Foreign Policy Analysis (FPA) regarding the behavior and decision-making processes within Greenland. The state/nation level of analysis will answer the question of how Greenland is navigating within IR and what kind of competencies and activities that Greenland has within foreign- and security policy. Here the theoretical framework will be concentrated around the concept of paradiplomacy. The systematic level of analysis is related to answering the questions of how far the self-government can be stretched or limited in relation to IR. This will be related to the theories of shelter theory and two-level games. Greenland will here be seen through cooperation with other countries and which countries that Greenland interacts with internationally. Can Greenland be seen as having some influence on a global scale?

## 1.5 Structure of the Book

The book will be divided into six subsequent parts after this introduction. First, the theoretical framework will be outlined with four various approaches used as lenses for the Greenlandic case: foreign policy analysis, paradiplomacy, shelter theory and two-level games theory. The four approaches are selected to give a more nuanced picture of Greenland's whereabouts. Foreign policy analysis can be seen as one of the traditional approaches within IR, while paradiplomacy and shelter theory have become more recent approaches related to small states and regions, however these can be seen as linked to foreign policy analysis as well. The two-level games theory is related to negotiations taking place at the international and the national level at the same time. Here various agreements that Greenland has entered will be analyzed. Second, methodological considerations will be discussed. Third, a historical background will be elucidated where the most important steps that Greenland has taken within international relations will be revealed. This will give the reader the background for understanding the Greenlandic context, and fourth the empirical mapping of the annual foreign policy reports and other important documents will be done. Fifth, the analysis will take place and thereafter a conclusion with an outlook will finalize the book.

## Notes

1 As of 6 April 2023, Pituffik/Thule Airbase changed its name to Pituffik Space Base.
2 The Arctic eight constitutes the following states: Canada, Denmark (incl. the Faroe Islands and Greenland), Finland, Iceland, Norway, Russia, Sweden, and the USA.
3 The cooperation within the Arctic Council is however for now only between the Arctic 7 without Russia, because of the Russian invasion towards Ukraine. Greenland has joined the EU's sanctions towards Russia, which has ended all seafood exports to Russia and the fishing cooperation is put on hold.
4 Departments and Ministries will be used interchangeably in this context, since during the evolvement Greenland has developed its structure and organization into ministries over time.

## References


Ackrén, Maria (2014). 'Greenlandic Paradiplomatic Relations'. In Lassi Heininen (ed.): *Security and Sovereignty in the North Atlantic*. Basingstoke and New York: Palgrave Macmillan, pp. 42–61.

Ackrén, Maria (2020). 'Grønlands rolle i internationale relationer', *Tidsskriftet Grønland*, 68:2, pp. 104–110.

Ackrén, Maria and Rasmus Leander Nielsen (2021). *The First Foreign- and Security Opinion Poll in Greenland*. Nuuk: Ilisimatusarfik/University of Greenland. Available at: https://uni.gl/media/6762444/fp-survey-2021-ilisimatusarfik.pdf

Ackrén, Maria and Uffe Jakobsen (2015). 'Greenland as a Self-governing Sub-national Territory in International Relations: Past, Current and Future Perspectives', *Polar Record*, 15:4, pp. 404–412.

Archer, Clive (1988). 'General Features of Political Development and Possibilities for Cooperation in the Arctic', *Current Research on Peace and Violence*, 11:4, pp. 137–145.

Åtland, Kristian (2008). 'Mikhail Gorbachev, the Murmansk Initiative, and the Desecuritization of Interstate Relations in the Arctic', *Cooperation and Conflict*, 43:3, pp. 289–311.

Friedheim, Robert L. (1988). 'The Regime of the Arctic – Distribution or Integrative Bargaining?', *Ocean Development and International Law*, 19:6, pp. 493–510.

*Greenland in Figures 2024*. Statistics Greenland. Available at: https://stat.gl/publ/en/GF/2024/pdf/Greenland%20in%20Figures%202024.pdf

*Greenland's Foreign, Security and Defense Policy 2024–2033*. Naalakkersuisut/Government of Greenland. Department of Statehood and Independence. Available at: https://paartoq.gl/wp-content/uploads/2024/03/Greenlands_Foreign_-Security_and_Defense_Policy_2024_2033.pdf

Heininen, Lassi and Heather N. Nicol (2007). 'The Importance of Northern Dimension Foreign Policies in Geopolitics of the Circumpolar North', *Geopolitics*, 12:1, pp. 133–165.

Henriksen, Alan K. (2020). 'The Arctic Peace Projection: From Cold War Fronts to Cooperative Fora'. In Gunhild Hoogensen Gjørv, Marc Lanteigne, and Horatio Sam-Aggrey (eds.): *Routledge Handbook of Arctic Security*. London and New York: Routledge, pp. 13–25.

Høegh, Kenneth (2024). 'Greenland's Foreign Policy, Past and Present: From the Merchants' Message to Bilateral Agreements', *Ocean and Coastal Law Journal*, 29:2, pp. 181–196.

Hoogensen Gjørv, Gunhild (2020). 'Understanding Arctic Security – What Has Changed? What Hasn't?'. In Gunhild Hoogensen Gjørv, Marc Lanteigne, and Horatio Sam-Aggrey (eds.): *Routledge Handbook of Arctic Security*. London and New York: Routledge, pp. 1–12.

Jacobsen, Marc, Ole Wæver, and Ulrik Pram Gad (2024). *Greenland in Arctic Security – (De)securitization Dynamics under Climate Thaw and Geopolitical Freeze*. Ann Arbor: University of Michigan Press.

Jacobsen, Marc and Ulrik Pram Gad (2018). 'Setting the Scene in Nuuk: Introducing the Cast of Characters in Greenland Foreign Policy Narratives'. In Kristian Søby Kristensen and Jon Rahbek-Clemmensen (eds.): *Greenland and the International Politics of a Changing Arctic – Postcolonial Paradiplomacy between High and Low Politics*. London and New York: Routledge, pp. 11–27.

Kleist, Mininnguaq (2019). 'Grønlands udenrigspolitik og internationale relationer: Nuværende rammer og mulig udvikling i et selvstændighedsperspektiv', *Politik*, 22:1, pp. 84–101.

Nature (2009). 'Arctic Black Gold Mapped'. Available at: https://www.nature.com/news/2009/090528/full/news.2009.527.html

Olsvig, Sara (2022a). 'Greenland's Ambiguous Action Space: Testing Internal and External Limitations between US and Danish Arctic Interests', *The Polar Journal*, 12:2, pp. 215–239. https://doi.org/10.1080/2154896X.2022.2137085

Olsvig, Sara (2022b). "Odd Couples' Win-Sets Maintaining U.S. Basing Rights through New Two-Level Game Negotiatiions with Greenland', *Scandinavian Military Studies*, 7:1, pp. 93–110. https://doi.org/10.31374/sjms.208

Padrtova, Barbora (2020). 'Applying Conventional Theoretical Approaches to the Arctic'. In Gunhild Hoogensen Gjørv, Marc Lanteigne, and Horatio Sam-Aggrey (eds.): *Routledge Handbook of Arctic Security*. London and New York: Routledge, pp. 29–42.

Pedersen, Torbjørn (2012). 'Debates over the Role of the Arctic Council', *Ocean Development & International Law*, 43:2, pp. 146–156.

Powell, Richard C. and Klaus Dodds (2014). 'Polar Geopolitics'. In Richard C. Powell and Klaus Dodds (eds.): *Polar Geopolitics? Knowledge, Resources and Legal Regimes*. Cheltenham and Northampton, MA: Edward Elgar Publishing, pp. 3–18.

Rahbek-Clemmensen, Jon and Camilla Tenna Nørup Sørensen (2021). 'Introduktion – stormagtskonkurrencen i Arktis og Nordatlanten og dens implikationer'. In Jon Rahbek-Clemmensen og Camilla T. N. Sørensen (eds.): *Sikkerhedspolitik i Arktis og Nordatlanten*. København: Djøf Forlag, pp. 17–37.

The Guardian (2011). 'Rush for Arctic's Resources Provokes Territorial Tussles'. Available at: https://www.theguardian.com/world/2011/jul/06/arctic-resources-territorial-dispute

US Geological Survey, USGS Fact Sheet 2008–3049. Available at: https://library.arcticportal.org/1554/1/usgs.pdf

Wegge, Njord (2011). 'The Political Order in the Arctic: Power Structures, Regimes and Influence', *Polar Record*, 47:241, pp. 165–176.

Young, Oran R. (1985). 'The Age of the Arctic', *Foreign Policy*, 61, Winter, pp. 160–179.

Young, Oran R. (2002). 'Can the Arctic Council and the Northern Forum Find Common Ground?', *Polar Record*, 38:207, pp. 289–296.

# 2 Theoretical Framework

Foreign policy is usually defined as the totality of a country's policies towards other countries and the interactions with foreign powers. This definition is quite broad and encompasses a variety of issue areas, including power, security, economic issues, environmental issues, human rights, population growth and migration, food, and energy policies, as well as foreign aid, development, and the relation between richer and poorer states (Breuning, 2007:5–6). Traditionally, the focus has been primarily on the quest to maintain and enhance a country's power and security. However, all that occurs between states and across nations is grounded in human decision-makers acting single-handed or in groups (Hudson, 2005). Decision-makers act according to their choices, decisions, and behavior. They almost always have several options to choose from regarding which decision to make and then act accordingly. The outcome can then be evaluated in the context when and where the decision was made (Breuning, 2007). To understand how, why, by whom, and on what basis decisions are made, as well as, how the contexts within which the decisions are made we need to understand the interplay between domestic and foreign policy (Breuning, 2007:9). Advisory systems and government bureaucracies always play some role in forming foreign policy decision-making and implementation. The world beyond the borders affects the possibilities for foreign policy action. These possibilities can present opportunities, but also constraints (Breuning, 2007:9).

In this book we will use three levels of analysis: the individual, the state/nation, and the international system level. The level-of-analysis approach was first introduced by Kenneth Waltz (Jackson and Sørensen, 2010). These three levels of analysis correspond to the different focus of foreign policy analysis: individuals decide which options to make and make the decisions, states/nations engage in foreign policy behaviors, and the interaction between states/nations in the international system leads to certain outcomes (Breuning, 2007:11). The connections between the three levels of analysis and their theoretical considerations are summarized in Table 2.1. This table has been inspired by Breuning's (2007) work but adapted to the Greenlandic context.

Inspired by the writings by Breuning (2007) the individual level of analysis focuses on leaders and decision-makers, the so-called elite. The focus here will be on the persons holding the office of Ministry for Statehood and Foreign Affairs

DOI: 10.4324/9781003387657-2

*Table 2.1* Levels of Analysis and Theoretical Considerations

| ***Levels of analysis*** | ***Foreign policy focus*** | ***Theoretical perspective*** |
|---|---|---|
| Individual | Options/decisions | Foreign policy analysis (FPA) |
| State/nation | Behaviors | Paradiplomacy |
| System | Outcomes | Shelter theory/two-level games |

and the officials working in the same department in the Greenlandic government, as well as the diplomats missioned abroad. Furthermore, the Premier of Greenland plays an overarching role as the signatory of relevant international documents and agreements. The assumption is that these individuals shape the course of history, because of their choices and decisions that drive the course of the events. The analysis might be directed towards their personalities or their perceptions. The personal traits, beliefs, and values can explain their foreign policy decisions and what motivates them to make certain decisions. The study of perceptions is more directed towards how the individuals view the world they are in. How are they framing the events or the situations they are experiencing? The individual level of analysis will depart from the foreign policy analysis (FPA) as the core theoretical umbrella. FPA is here seen as the overall theoretical lens, but other theoretical lenses might also be useful. The individual level of analysis is related to the objectives and aims regarding the foreign- and security policy that Greenland has on an overarching basis.

The state/nation level of analysis focuses on factors related to the domestic area of the state/nation. This type of analysis includes the institutional framework of the state/nation. This means the relationships between the executive and legislative branches of government, the organization of the government bureaucracy, or whether the state is a democracy, domestic entities (such as interest groups, ethnic groups, or public opinion more generally), economic conditions, and the state's/nation's history and culture. The analysis will shed light on how the Greenlandic political system works and how Greenland behaves within foreign policy. This will constitute a major part of the analysis in the book related to the content of official documents and results from recent opinion polls undertaken (see Nielsen and Ackrén, 2021 and Nielsen and Ackrén, 2024). The emphasis will here be on the paradiplomatic framework as the core theoretical lens, which is linked to foreign policy analysis. This analysis takes the behavior of Greenland into account and how Greenland navigates, and in which circumstances Greenland acts as its own nation within foreign policy. This is also related to the channels used by the Greenlandic government and room for maneuver within IR.

The system level of analysis focuses on the comparison between states/nations. This level of analysis encompasses the relative strength of states/nations. The international system is defined as a set of states/nations whose interactions are formed by their relative strength, such as their power and wealth. These capabilities influence their possibilities for action and success on the global stage. In this book we use the system level of analysis as the focus of the interactions Greenland has with Denmark, USA, the EU, and other international organizations. The theoretical lenses to be used here are the shelter theory and the two-level game. The shelter theory is used to see if Greenland is balancing towards the US or Denmark as having these states as its shelter in foreign policy or if Greenland is seeking shelter from somewhere else. The two-level game is relevant, since Greenland sometimes must satisfy both the domestic constituencies and international imperatives simultaneously, which can become a delicate balancing act. This situation can especially become intriguing when the domestic and international environment push decision-makers in different directions. Here various scenarios will be exemplified for how far Greenland can go or how some limitations might be posed by Denmark. The theoretical lenses will be described more thoroughly in the next paragraphs.

## 2.1 Foreign Policy Analysis

Foreign policy analysis has often focused on state behavior and how states react on external and internal pressures. Both external and internal pressures are affecting decision-making within the area of foreign policy. There are different approaches that can be discern theoretically: power relations, decision-making processes, and input-output analyses. Models often include variables such as power input, power output, interests, and objectives (Brecher et al., 1969). These are linked to how states act within the area of foreign policy. Every state has its own interests and objectives in terms of how to outline and make their voices heard internationally. There are many various channels in ways of being part of international organizations, having bilateral or multilateral treaties with other countries, through diplomatic missions and so forth. A variety of different actors are involved at different levels within the state apparatus. Foreign policy involves goals, strategies, measures, methods, guidelines, directives, understandings, and agreements by which national governments conduct international relations with various actors (Jackson and Sørensen, 2010).

In the first period of FPA scholarship the focus was on conceptualization and the development of actor-specific theory at various levels of analysis. There was also an emergence of examining the influence of organizational process and bureaucratic politics on foreign policy decision-making (Hudson, 2005). Later, the comparative foreign policy became the focus where cross-national and multilevel theory was the subject on the agenda. The interest became to answer the questions of "who does what to whom and how" in international affairs (Hudson, 2005). A traditional approach to FPA was also about government's external policies in terms of knowing the history or background, comprehending the interests and concerns driving the policies. This included the knowing of the outcomes and consequences of past foreign policy decisions and actions. The ability to recognize the circumstances under which the government operates carrying out its foreign policy was also of importance (Jackson and Sørensen, 2010).

The state-centric perspective has become seen as outdated as new actors have come into the scene and as new forces have altered the nature of international relations by connecting states into a network of interdependencies (Smith, 1986). This is especially true of the integration of countries into trade associations or economic blocks, or the EU as a large political project where states have transferred some of their sovereignty to a supranational level. International organizations have become important actors of IR as well with their own agendas and motives. Foreign Policy Analysis (FPA) has been criticized for being focused on the political-military context, but since the 1980s also economic factors have become central in FPA. Most of the work is concentrated on single case studies and therefore FPA has not reached the level of becoming a general theory (Smith, 1986). Rational choice theory is usually used in connection with FPA, since this focuses on the ability of individualistic maximizing of self-interest by the actors. There are also various forms of game theory, which have been used to illustrate states' actions in different scenarios (Carlsnaes, 1992).

Theories of international relations are usually based on assumptions that states are unitary actors with a motive or wish to survive. In this sense, the theories then try to explain how variations in conditions external to states push or pull them in various directions. A theory of foreign policy tries to explain why states behave in different ways in a world system. Differences in behavior arise from differences in internal composition. Foreign policies become governmental products, where the performance of governments is in focus (Waltz, 1996). Neoclassical realists stress that foreign policy decisions are made by individuals, political leaders, and elites. The decisions are influenced by their perceptions of relative power, perceptions of intent, state motivations and how the values of the leaders interact with their cost-benefit analysis (Wivel, 2005). Therefore, the scope and ambition of a country's foreign policy is driven by its place in the international system and specifically by its relative material power capabilities (Wivel, 2005). Those who are more concerned with multilateral questions tend to take a more liberal view, emphasizing international institutions as means of reducing international conflict and promoting mutual understanding and common interests (Jackson and Sørensen, 2010).

The decision-maker's mind can be seen as a microcosm where various factors such as culture, history, geography, economics, political institutions, ideology, demographics, and so on shape the societal context in which the decision-maker operates. The individualistic traits are also complex including beliefs, attitudes, values, experiences, emotions, style, memory, national and self-conceptions which relates to the behavior of the decision-maker (Hudson, 2005). Personality of individual decision-makers, organizational setting and bargaining among groups are all interrelated and mixed in forming foreign policy outcomes. In the end it is foreign policy makers who make decisions about how a country/nation should behave internationally. These elites behave on their idea of what would be acceptable for the people (Cantir and Kaarbo, 2012). Research in FPA suggests that elites and masses might disagree on their country's/nation's proper roles in the international system. There is usually a disconnection between leaders' and public's view on several specific issues (Cantir and Kaarbo, 2012). We must remember that many foreign policy decisions and actions are taken in uncertain contexts and with imperfect information. Foreign policy is more prone to uncertainty and more exposed to instability and conflict than domestic policy (Jackson and Sørensen,

2010). Domestic policy is carried out under the jurisdiction of a sovereign government, which possesses the legal authority to preside over the domestic policy. In the current international scene, uncertainty has become even more prominent with Russia's invasion of Ukraine, Israel's war against Hamas and Palestine, cyberattacks over internet, which affects governments and critical institutions (e.g. health sector, hospitals, companies, elections, etc.) and hybrid warfare with disruptions of infrastructure, such as underwater cables and energy supply and so forth.

The individual level of analysis will map the relevant people involved within the foreign policy domain within the Greenlandic government from 2003 to the latest period of 2023. The idea is to map what kind of individuals that have been represented in the Greenlandic government, what the objectives have been within IR and what kind of outcomes that have come out during the different periods in form of international agreements and participation within international organizations and so forth. The people will be set into the historical context of their time since this might have been a factor in why certain decisions and actions have been made.

## 2.2 Paradiplomacy

Paradiplomacy has become a concept used for defining regions' possibilities to act within IR, such as subnational governments' activities within IR. Subnational governments perform activities within IR in various ways: they open trade and cultural missions abroad, sign treaties and agreements with foreign powers and non-state actors, they participate in international networks of regional cooperation and they even sometimes challenge the official foreign policy of their central governments through their statements and actions (Kuznetsov, 2015). Paradiplomacy is an activity that falls into a legal and constitutional grey zone, since foreign powers typically are issues of state affairs (Lecours, 2008). Recent decades have seen an explosion of international activity in the part of regions, stateless nations across the globe, but especially in a European context. For example, there are over 200 regional offices in Brussels, lobbying the European Commission, networking with each other, and getting involved in the European policy communities (Keating, 2000).

Paradiplomacy occurred in the 1970s and since then it has been flourishing and been seen as an important issue within IR. During the decades paradiplomacy has been investigated through various dimensions, such as, constitutional, nationalistic, area/border, regionalization/globalization security/geopolitical and global economy dimensions (Kuznetsov, 2015). The research has mainly focused on federal states and regions in Europe. Paradiplomacy is about the behavior of subnational entities when they encounter the international sphere. It is about how the subnational governments act and behave towards foreign actors and which channels they use in order to influence their own foreign policy agenda. There might be different motives for regions to act internationally. They might be strategically located and equipped with natural resources, they might have some niche markets or products which are of global interest, they might be part of various international networks and international organizations and so forth.

Criekemans (2008) has identified three waves of paradiplomacy. The first wave was taken place in the 1980s when regions with legislative competencies were trying to attract foreign direct investments to their regions or using culture and identity as a marker to place themselves on the international scene. These initiatives were usually of an ad hoc nature. The second wave occurred during the 1990s in relation to the integration of Europe, where regions established more jurisdictional bases for their international involvement by legislative reforms or establishment of foreign offices abroad. The third wave is characterized by a verticalization of the organizational structure of the administration into foreign departments or offices of foreign affairs at the regional authority level.

The jurisdictional capacities of subnational governments vary, and their degree of self-government differs, but those regions and territories which have an aim to become independent or receive more autonomy might have certain interests in using paradiplomacy as a channel to achieve their goals (Grydehøj, 2014). This puts pressure on the central governments they are part of as well and therefore competing interests might come at stake.

Subnational governments do not act in a vacuum and the tendencies of globalization, and the rise of transnational regimes have somewhat eroded the distinction between domestic and foreign policy, which has changed the responsibilities between states and subnational entities. Paradiplomacy has become a field of international activity which simulates or approximates official and conventional international relations (Bartmann, 2006).

André Lecours (2008) has identified three layers where paradiplomacy is used actively to promote identity and interests: economic issues, cooperation, and political considerations. Economic issues cover the aim of attracting direct foreign investments towards the region and developing strategies for international companies to become established in the area. This also goes hand in hand with targeting new markets and products for export. Cooperation can have many dimensions since cooperation can be done at different governmental levels and between various stakeholders. It covers areas such as culture, education, technical matters, and others. This usually takes the form of transborder initiatives as for example the "Four Motors in Europe" or the Assembly of European Regions. The cooperation contains development assistance, cultural and educational exchanges. The third layer of political consideration is then about the region's effort to promote itself internationally through diplomatic exchanges to seek influence on the global stage. This contains the aspects of being part of international organizations, establishing mini embassies abroad or signing bilateral or other international agreements.

According to Paquin and Lapachelle (2005) three causes of why paradiplomacy has entered the scene within international relations can be elucidated: the nation-state crisis and globalization, nationalism, and internationalization processes. Globalization has driven regions and large metropolitan areas for the acquisition of shares in the world market. Multinational corporations are attracted to regions outside the capitals. Nationalism has given fuel to separatist and minority movements and especially within regions where various minorities act this can give rise to international strategies in search of resources or support that are unavailable to them within their own countries. Another aspect is to connect with diasporas outside the borders and in this way define their own international interests and needs. The internationalization processes refer to international agreements and treaties, where issues such as environment, health, communication, social services, transportation, use of land and other issues need to be implemented on the regional, subnational level. Many European countries have decentralized systems, where a lot of such issues are handled by the subnational governments.

Paradiplomacy is a complex process that operates at different levels, from media events of political visits to the day-to-day matters those local officials are undertaking. There are both successes and failures, where regions emerge at transnational scenes (Keating, 2000). Indigenous peoples in the Arctic have been engaged in internal, but also external diplomacy for decades. Within the federal system of Canada, for instance, treaty-making and negotiations have been taken place between the diverse levels of authority, from the federal level to the regional or provincial level and with indigenous organizations. The external more international diplomatic relations have been between state and non-state actors. The Inuit Circumpolar Conference (established in 1977, now called Inuit Circumpolar Council) is an example of transborder cooperation between four nations: Canada, Alaska, Greenland, and Russia, where Inuit land rights, human rights, and environmental protection have been in focus. Inuit Circumpolar Council (ICC) became acknowledged by the United Nations in 1983 and became granted the status of non-governmental organization. In this way ICC is working closely with other indigenous organizations as part of the UN Permanent Forum on Indigenous Issues (UNPFII) (Abele and Rodon, 2007).

More recent research has tried to develop the paradiplomatic theoretical framework and one new article by Kovacikova (2024) is looking at paradiplomacy through a new data set with sub-state governments from all OECD countries. She is using 517 sub-state governments in 2019 to show why some sub-state governments enter international relations, while others do not (Kovacikova, 2024). To test this, she uses capacity-based hypotheses with three overarching variables of revenue, human capital and information (divided into geographic and economic information). The results show that only 17% of all sub-state governments have an active paradiplomatic policy within the OECD countries. It is also concluded that greater economic, human capital, and information capacities are more likely to pursue paradiplomacy as their part of international economic development strategy. Revenue is stated as the most important driver for paradiplomatic activities (Kovacikova, 2024:20).

Looking at the Greenlandic case we see that Greenland has developed from a colony towards an extended self-governing region, which nowadays also includes international relations. Greenland acts as an own nation within areas that have been transferred from the Danish authorities to the Greenlandic authorities. Greenland will here be described and analyzed in Chapter 4 with the historical background and the relationships towards foreign powers and international involvement in international organizations and in Chapter 5 the activities of the Government of Greenland will be outlined. This is the most detailed and described case study done so far in relation to Greenland's role within IR and how paradiplomacy is used in the Greenlandic case.

## 2.3 Shelter Theory and Two-Level Games

### *2.3.1 Shelter Theory*

Shelter theory has become a theory directed towards small states and self-government areas. Small states are seen as vulnerable actors within IR, and this also relates to self-government areas. To have a voice within IR small states and self-governing areas might lean towards a larger power to seek shelter and to be able to make their voices heard at the international arena (Wivel et al., 2014; Thorallsson et al., 2019). To follow a larger state or team up with a larger power might be a way to receive some benefits on the international scene in form of protection of potential hostile attacks from foreign powers or to benefit culturally, economically or receive possibilities for exchange programs in the educational sphere for instance (Thorallsson et al., 2019). The capacities of small states and small regions are often marginal when it comes to security and foreign power relations. Some small states might even be without a military, such as, Iceland. Iceland has therefore become a NATO member to be able to protect its borders and citizens. This means, of course, that the country is very dependent on the goodwill from outside powers. The USA has been the protector of Iceland for many decades, but in 2006 the US military withdraw their troops from the Keflavik airbase, however, recently a new effort to rebuild the base has become topical due to the enhanced activities taken place in the North (Thorallsson et al., 2019). The Keflavik base is opened again with NATO countries interchangeably present surveilling the area around Iceland.

Small states or regions need to compensate for their inbuilt structural weaknesses and smallness and therefore aim to lean towards larger states or international organizations for protection. Small entities have smaller, or no military powers and therefore might pursue defensive alliances to enhance their security and power by pooling resources (Thorallsson and Steinsson, 2019:26–27; Thorallsson and Steinsson, 2018). Small states and regions might need political, economic, and societal shelter. These forms of shelter might be provided by larger states and international organizations. The political shelter takes the form of direct and visible diplomatic or military security offered by a larger state or an international organization. Economic shelter includes direct economic assistance or help from an external financial institution, a currency union or access to a common market, which is provided for by a more powerful state or by an international organization. Societal shelter involves diffusion of foreign people and ideas to avoid social stagnation or to receive more knowledge from other states or international organizations (Thorallsson and Steinsson, 2019; Thorallsson, 2019).

Shelter theory highlights three interrelated issues of concerns to small states and regions: the reduction of risk before an eventual crisis; assistance in absorbing shocks when a risk or crisis occur; and assistance of recovering in the aftermath of a crisis (Thorallsson and Steinsson, 2019). Shelter theory provides a framework for analyzing the costs and benefits associated with small states or regions relation to both their domestic national level, as well as external relations (Thorallsson and Steinsson, 2018). In this way, shelter theory becomes a form for risk management within IR. Usually, small states or regions tend to choose shelter from like-minded states and organizations. For example, Greenland has sought shelter from Denmark in over 300 years in political, societal and economic terms, and the US has been a military/security protector since the Second World War, and the Nordic countries are seen as societal shelter with similar political, social, and cultural systems. Chapter 4 will illustrate the form of shelter Greenland has chosen over time. Shelter theory will be used to describe various scenarios where Greenland seek shelter for different purposes.

### *2.3.2 Two-Level Games*

The theory of two-level games has been addressed by Putnam (1988). The idea is that we cannot totally distinguish domestic and foreign policy from each other. The levels of politics are not two pillars which are totally working in isolation. Negotiations and interests are proposed from both levels. At the national level we have politicians and interest groups affecting the political agenda and pressure the government to adopt favorable policies. At the international level, national governments seek to maximize their efforts to satisfy domestic pressures, while at the same time minimizing any counter consequences of foreign developments. The two games are interrelated and cannot be ignored by decision-makers. Each national political leader takes part in both games (Putnam, 1988:434).

The bargaining taking place at the national level, called Level 1, usually leads to a tentative agreement. At the international level, called Level 2, separate discussions within each group of negotiators are then leading to whether the agreement should be ratified or not (Putnam, 1988:436). The ratification procedure clearly affects the outcome. For example, if a two-thirds vote is required for ratification, the preferable outcome will certainly be less likely than if only a simple majority is needed (Putnam, 1988:448). However, not all significant ratification procedures are formalized. The ratification of agreements might also be based on informal practices. Strong discipline within the government party, for example, increases the likelihood for positive outcomes where the Level 1 negotiator can expect backing from government or parliament (Putnam, 1988:449). The logic of the two-level game derives from the assumption that the chief negotiator is the only formal link between Level 1 and Level 2 in this context. The motives of the chief negotiator include enhancing his/her standing in the Level 2 game by increasing or minimizing potential losses. For example, a Head of Government may seek popularity if he/she concludes a successful international agreement, or he/she may anticipate that the results of the agreement will be politically rewarding. Another motive may be to shift the balance of power at Level 2 in favor of domestic policies. A third motive may be to pursue his/her conception of the national interest in the international context (Putnam, 1988:456–457).

The Level 2 games function simultaneously on both levels. The government representatives deal with negotiations internationally and interactively with their counterparts, and at the same time they need to negotiate with their constituents at the domestic level. Agreements at the international level is only possible when the win-sets of the involved parties overlap (da Conceição-Heldt and Mello, 2017). Domestic political institutions are important in order to explain foreign policy. These institutions include the role interest groups play affecting the government representatives, power-sharing mechanisms between the executive and the legislature are also important, and the number of eventual veto players might also play a role (da Conceição-Heldt and Mello, 2017).

Greenland is playing two-level games with Denmark in various contexts and especially in foreign policy. Nowadays, also two-level games are played with the USA in the form of Memorandum of Understandings (MoUs) for instance. These two-level games will be illustrated by recent examples in a specific section in Chapter 4 in the book.

## 2.4 Summary

As illustrated in Chapter 1 with Figure 1.1 we will combine the three levels of analysis with the four theoretical frameworks to be able to answer the research questions. Greenland is interacting within IR at various levels. At the individual level we have the decision-makers within the institutional structure of self-government. Here the objectives and aims are formulated within IR and which strategy Greenland will follow in relation to the outside world. The state/nation level has to do with the channels used and how Greenland is navigating within IR. This is also somewhat related to the capacities and resources that Greenland has regarding the area of foreign- and security policy. The international systematic level of analysis is how far Greenland can stretch its autonomy and which limitations there might be in relation to Denmark regarding IR. The systematic level is also related to the cooperation and network Greenland has in foreign- and security policy. The systematic level will also place Greenland into a more global context.

## References

Abele, Frances and Thierry Rodon (2007). 'Inuit Diplomacy in the Global ERA: The Strengths of Multilateral Internationalism', *Canadian Foreign Policy Journal*, 13:3, pp. 45–63.

Bartmann, Barry (2006). 'In or Out: Sub-national Island Jurisdictions and the Antechamber of Para-diplomacy', *The Round Table: The Commonwealth Journal of International Affairs*, 95:386, pp. 541–559.

Brecher, Michael, Blema Steinberg, and Janice Stein (1969). 'A Framework for Research on Foreign Policy Behavior', *Journal of Conflict Resolution*, 13:75, pp. 75–94.

Breuning, Marijke (2007). *Foreign Policy Analysis. A Comparative Introduction*. New York: Palgrave Macmillan.

Cantir, Christian and Juliet Kaarbo (2012). 'Contested Roles and Domestic Politics: Reflections on Role Theory in Foreign Policy Analysis and IR Theory', *Foreign Policy Analysis*, 8, pp. 5–24.

Carlsnæs, Walter (1992). 'The Agency-Structure Problem in Foreign Policy Analysis', *International Studies Quarterly*, 36:3, pp. 245–270.
Criekemans, David (2008). 'Are the Boundaries between Paradiplomacy and Diplomacy Watering Down? Preliminary Findings and Hypotheses from a Comparative Study of Some Regions with Legislative Power and Small States', Paper presented at the 2nd Global International Studies Conference: *What Keeps Us Apart, What Keeps Us Together? International Order, Justice and Value*, Faculty of Social Sciences, University of Ljubljana, Slovenia, 23–26 July 2008.
da Conceição-Heldt, Eugénia and Patrick A. Mello (2017). 'Two-Level Games in Foreign Policy Analysis'. In William R. Thompson (editor in chief): *Oxford Research Encyclopedia of Politics*. New York and Oxford: Oxford University Press, pp. 770–789. https://doi.org/10.1093/acrefore/9780190228637.013.496
Grydehøj, Adam (2014). 'Goals, Capabilities, and Instruments of Paradiplomacy by Subnational Jurisdictions'. In Adam Grydehøj, Linda Fabiani, Jordi Solé Ferrando, Lorena Lopez de Lacalle Aristi, and Maria Ackrén (eds.): *Paradiplomacy*. Brussels: Centre Maurits Coppieters, pp. 10–21.
Hudson, Valerie M. (2005). 'Foreign Policy Analysis: Actor-Specific Theory and the Ground of International Relations', *Foreign Policy Analysis*, 1, pp. 1–30.
Jackson, Robert and Georg Sørensen (2010). *Introduction to International Relations: Theories and Approaches*, 4th Edition. Oxford: Oxford University Press.
Keating, Michael (2000). 'Paradiplomacy and Regional Networking'. In *Forum of Federations*. Ottawa: Canada.
Kovacikova, Lucia (2024). 'Bringing Capacity Back In: Paradiplomacy and Sub-State Government Internationalization in the Global Economy', *Publius: The Journal of Federalism*, 3rd October 2024, 55:1, pp. 1–27. https://doi.org/10.1093/publius/pjae039
Kuznetsov, Alexander S. (2015). *Theory and Practice of Paradiplomacy – Subnational Governments in International Affairs*. London and New York: Routledge.
Lecours, André (2008). 'Political Issues of Paradiplomacy: Lessons from the Developed World'. In Virginie Duthoit & Ellen Huijgh (eds.): *Discussion Papers in Diplomacy*. Netherlands Institute of International Relations 'Clingendæl': Antwerp University, pp. 1–22.
Nielsen, Rasmus L. and Maria Ackrén (2021). 'Grønlændernes holdninger til udenrigs- og sikkerhedspolitiske spørgsmål: indsigter fra en survey', *Økonomi & Politik*, 94:2, pp. 84–98.
Nielsen, Rasmus L. and Maria Ackrén (2024). *The Second Foreign and Security Policy Opinion Poll in Greenland*. Nuuk: Nasiffik/Ilisimatusarfik (University of Greenland). Available at: https://da.uni.gl/media/4bwb5ict/survey-report-leander_ackre-n_final.pdf
Paquin, Stéphane and Guy Lapachelle (2005). 'Why Do Sub-States and Regions Practice International Relations?'. In Guy Lapachelle and Stéphane Paquin (eds.): *Mastering Globalization: New Sub-states' Governance and Strategies*. London and New York: Routledge, pp. 77–89.
Putnam, Robert D. (1988). 'Diplomacy and Domestic Politics: The Logic of Two-level Games', *International Organization*, 42:3, pp. 427–460.
Smith, Steve (1986). 'Theories of Foreign Policy: An Historical Overview', *Review of International Studies*, 12:1, pp. 13–29.
Thorallsson, Baldur (2019). 'Shelter Theory and Iceland', *Middle East Insights*, no. 128, 18 November 2019.
Thorallsson, Baldur and Sverrir Steinsson (2018). 'A Theory of Shelter', *Policy Brief*, no. 1, 26 June 2018, Small States and the New Security Environment (SSANSE), Research Project Policy Briefs.

Thorallsson, Baldur and Sverrir Steinsson (2019). 'A Theory of Shelter'. In Baldur Thorallsson (ed.): *Small States and Shelter Theory – Iceland's External Affairs*. London and New York: Routledge, pp. 24–48.

Thorallsson, Baldur, Sverrir Steinsson, and Thorstein Kristinsson (2019). 'The Small State in International Relations'. In Baldur Thorallsson (ed.): *Small States and Shelter Theory – Iceland's External Affairs*. London and New York: Routledge, pp. 1–23.

Waltz, Kenneth N. (1996). 'International Politics Is Not Foreign Policy', *Security Studies*, 6:1, pp. 54–57.

Wivel, Anders (2005). 'Explaining Why State X Made a Certain Move Last Tuesday: The Promise and Limitations of Realist Foreign Policy', *Journal of International Relations and Development*, 8, pp. 355–380.

Wivel, Anders, Alyson J. K. Bailes, and Clive Archer (2014). 'Setting the Scene: Small States and International Security'. In Clive Archer, Alyson J. K. Bailes, and Anders Wivel (eds.): *Small States and International Security: Europe and Beyond*. London and New York: Routledge, pp. 3–25.

# 3 Methodology

This chapter discusses the methodology used in this book. The overall method used is document and content analysis, as well as a case study approach, since the book is mapping documents of various kinds and is only looking at one single case, i.e. Greenland.

## 3.1 Document and Content Analysis

Document and content analysis are one of the most used methods within social sciences. It is almost impossible to avoid empirical investigations without involving documents in one way or another. Document analysis is a systematic approach to analyzing a certain amount of information. Often document analysis focuses on a specific period of development to identify stability and changes in each area (Lynggaard, 2020:185). This can be related to policy processes, a discursive field, or a historical development. There are usually four criteria to have in mind when analyzing documents: (1) *authenticity* (is it genuine, authentic, and original?), (2) *credibility* (are the contents free from error and distortion?), (3) *representativeness* (is it typical of what it represents?), and (4) *meaning* (is the meaning comprehensible?) (Bryman et al., 2022:243–244). Lynggaard (2020) writes about three forms of documents, which can be diversified. First, we have the primary sources of information, where the information has been circulated among a closed group of people at a specific time. This kind of document has been near the event or the situation where the document was referred to. These types of documents are, for example, minutes from meetings, personal letters, or presentations of negotiations internally, which have not been outlined in public. Second, there are secondary documents in the form of public documents, which everyone can get hold on. These documents are typically legal texts, government reports, newspaper articles or declarations of intent from companies, interest organizations and political parties. Thirdly, the third kind of documents are, as secondary documents open access for all but have been produced after an event or situation has occurred. These documents are typically academic books and articles (Lynggaard, 2020:187–188). Other authors divide the documents into written and un-written forms of documents, where we have three written forms of documents with personal documents (for example diaries, letters, e-mails, transcribed interviews, biographies, personal blogs, etc.), internal

DOI: 10.4324/9781003387657-3

organizational documents (such as minutes from meetings, internal investigations, consulting reports, project descriptions etc.), public documents (for example public statistics, government reports, investigations, legal texts, books and articles and so on), while the fourth category is un-written forms of documentation in form of television, You Tube clips, photographs, paintings, diagrams, graphs, etc. (Lodsgaard, 2021:201–211).

The idea behind document and content analysis is to limit the investigation to take a sample of documents into account. These documents are then treated as data. The data can be analyzed through both quantitative and qualitative approaches. The quantitative approach collects numerical data on specific elements found in the documents to study frequencies and patterns of these elements statistically. This kind of analysis is used to emphasize how frequent certain images, words, or ideas are stated in relation to other images, words, or ideas. The qualitative approach examines both the content and the contextual meaning of communication. This is a more inductive way of dealing with the data to find themes and patterns of interest. The researcher will in this sense use a clear strategy and rules for categorizing the material in the form of coding (Bryman et al., 2022:252, 254). Usually, both quantitative and qualitative approaches are used in combination within the study of interest.

The meaning of a text can be identified by the sender or producer of the text and by the receiver or consumer of the text. The interpretation is done between the sender and the receiver. It can sometimes be hard to interpret what the sender/author has meant with the text. The receiver/reader of the text can also interpret things differently (Lindkvist, 1981:23–25). Within content analysis it is common to distinguish between analysis and interpretation. The analysis consists of different forms of treatment of the content of the text, for example, coding and counting frequencies of words or thematic structures of the text, while interpretation is done in part or as a whole and linked to the analysis. It is therefore both the elements and the structure of a text that is in focus (Lindkvist, 1981:27–28).

The text or content analysis is furthermore research question oriented (Mayring, 2019). The main aim of the investigation is therefore to answer the research questions posed at the beginning of the study. The questions should be answered at the end of the analysis. The qualitative content analysis is done in two major phases. The first phase is related to data generation where the researcher is absorbing the content and determining the unit of analysis and thereafter develops coding schemes and undertakes the preliminary coding of the content. In the second phase the researcher is establishing categories across the coding schemes and is trying to identify themes/patterns across the categories and thereafter draw the interpretations and implications of the analysis (Roller, 2019). The document and content analysis should cover three expectations that the researcher has: the theoretical conceptions one has about the field of research, the concrete empirical work one has knowledge about, and the general conceptions about the linkages between theory and empirical frameworks one has beforehand (Elklit and Jensen, 2020:153).

The usage of annual foreign policy reports as the main analyses of documents within this study will give the reader a description of what the Government of

Greenland has done throughout the 20 years within the sphere of foreign- and security policy. This will also shed light on which the priorities have been and what kind of activities that the Government has been undertaking. The idea is also to give the reader a chronological order of the events as far as possible. All foreign policy reports are public and therefore easily available. At the same time, they represent authentic and credible forms of material. Furthermore, other sources of material are used, such as academic books and articles, legal documents, other reports and news articles. This is to complement the research and give a broader picture of the foreign- and security policy in Greenland.

## 3.2 Case Study

A case study is usually a spatially delimited phenomenon (a unit) observed at either a single point in time or over a period. A case study consists of a detailed investigation that provides an analysis of the context and processes involved in the phenomena under investigation (Meyer, 2001:329). The rather loose design implies that there are several ways that need to be addressed regarding how the research is conducted. The approach is usually useful to answer *what*, *how* and *why* questions in research (Meyer, 2001:330). In principle, every study is a case study because it is an analysis of social phenomena specific to time and place (Ragin, 2005:2).

There are three overarching strategies to be considered in relation to case studies. First, there is the analytical strategy, where the researcher looks at the world objectively and where specific events might decide the outcomes. In this context, the researcher is looking for causal relationships between dependent and independent variables. Secondly, we have the systematic strategy where the reality is seen to exist independently of the investigation in question. Here the world is addressed as systems, which consist of components and their relationships towards each other. Thirdly, we have the actor strategy where reality is perceived as a social construction. The world is established by the actors' activities, language and behavior. This is a more subjective way to deal with a case study (Evald and Freytag, 2021:116–117).

The case study will be used to test the theoretical lenses that have been described in Chapter 2 and therefore the case study will link these theoretical frameworks in relation to the three levels of analysis, i.e. the individual, the national and the systematic level. The case study will be an in-depth study of Greenland as an actor within the foreign- and security policy domain and at the same time the process of development will be scrutinized over time. Greenland can also be seen as a within-case study since we are looking at various aspects of entities within the case. We are using three different levels of analysis with individuals, the Government and the external context with other states and organizations, so therefore Greenland becomes not only a single entity as such since we also are looking at the case from a broader perspective and throughout a period of 20 years. This kind of case study is also called a longitudinal or diachronic case study (see Gerring, 2008).

In the study by Jacobsen, Wæver and Gad (2024) Greenland is featured as a unique case study. It is not stated as such, but between the lines it is possible to get that impression. However, in this book we are careful to use Greenland as a unique case, since it would require a comparative study initially to be able to say if that is the case. If no comprehensive comparative works have been done in this area it is hard to state that Greenland features a unique case.

## 3.3 Benefits and Drawbacks with Methods Used

The benefits with the document and content analysis are that the researcher has access to the documentation under study and that these documents are relevant to answering the research questions. The case study is designed to give the researcher flexibility to investigate the case from diverse perspectives. The idea is to seek and describe and make sense of the phenomenon and strive for objectivity and intersubjectivity. Intersubjectivity relates to whether other researchers can trace the interpretations made in the case and that it is possible to follow the general methods and procedures of the analysis (Meyer, 2001:344–345).

Validity refers to whether the evidence the theoretical framework correctly corresponds with the observations done. The strength of qualitative research lies in the flexible and responsive interaction between researcher and data. The advantage is that it is possible to explore the concepts in more detail and the relationships between them. In this way multiple sources are used to build up constructs and measurements and multiple viewpoints (Meyer, 2001:345–346).

The drawbacks with the above selection of methods are that the researcher might have interpreted the documents in a way that others might not have done. The interpretation of the data within the documents for the analysis can be done in many ways. This lies totally in the hands of the researcher, while it is therefore important to describe the process of analysis in detail. Another major drawback is that it will be impossible to come up with a generalizing view of how subnational regions act within foreign- and security policy, since we have only used one single case. However, by providing a richer and deeper look into the case, it allows the readers to make their own judgements about the applicability of the case in other contexts. A third drawback is related to reliability. Reliability focuses on whether the process of the study is consistent and reasonable stable over time and that other researchers can replicate the study. However, it might not be surprising that different researchers generate different results and reach different conclusions. This can only be controlled by being open-minded about the departure and the framework used as well as how the procedures regarding collection of data and analysis are done.

## References

Bryman, Alan, Edward Bell, Jennifer Reck, and Jessica Fields (eds.) (2022). *Social Research Methods*. New York and Oxford: Oxford University Press.

Elklit, Jørgen and Henrik Jensen (2020). 'Kvalitative Datakilder'. In Lotte Bøgh Andersen, Kasper Møller Hasen and Sune Welling Hansen (eds.): *Metoder i statskundskab*. København: Hans Reitzels Forlag, pp. 151–179.

Evald, Majbritt and Per Vagn Freytag (2021). 'Casestudiet – forskellige formål og anvendelser'. In Søren Voxted (ed.): *Valg der skaber viden – om samfundsvidenskabelige metoder*. København: Hans Reitzels Forlag, pp. 115–135.

Gerring, John (2008). *Case Study Research – Principles and Practices*. Cambridge: Cambridge University Press.

Jacobsen, Marc, Ole Wæver, and Ulrik Pram Gad (2024). *Greenland in Arctic Security – (De) securitization Dynamics under Climate Thaw and Geopolitical Freeze*. Ann Arbor: University of Michigan Press.

Lindkvist, Kent (1981). 'Approaches to Textual Analysis'. In Karl Erik Rosengren (ed.): *Advances in Content Analysis*. Sage Annual Reviews of Communication Research, vol. 9. Beverly Hills, CA and London: Sage Publications, pp. 23–41.

Lodsgaard, Lise (2021). 'Dokumenter – en overset datakilde'. In Søren Voxted (ed.): *Valg der skaber viden – om samfundsvidenskabelige metoder*. 2. udgave. København: Hans Reitzels Forlag, pp. 209–223.

Lynggaard, Kennet (2020). 'Dokumentanalyse'. In Svend Brinkmann and Lene Tanggaard (eds.): *Kvalitative metoder – En grundbog*. 3. udgave. Hans Reitzels Forlag, pp. 185–202.

Mayring, Philipp (2019). 'Qualitative Content Analysis: Demarcation, Varieties, Developments', *Forum Qualitative Sozialforschung/Forum Qualitative Social Research*, 20:3, Art. 16. https://doi.org/10.17169/fqs-20.3.3343

Meyer, Christine B. (2001). 'A Case in Case Study Methodology', *Field Methods*, 13:4, pp. 329–352.

Ragin, Charles C. (2005). 'Introduction: Cases of "What Is a Case?". In Charles C. Ragin and Howard S. Becker (eds.): *What Is a Case? Exploring the Foundations of Social Inquiry*. Cambridge: Cambridge University Press, pp. 1–21.

Roller, Margaret R. (2019). 'A Quality Approach to Qualitative Content Analysis: Similarities and Differences Compared to Other Qualitative Methods', *Forum Qualitative Sozialforschung/Forum Qualitative Research*, 20:3, Art. 31. https://doi.org/10.17169/fqs-20.3.3385

# 4 Background of Greenland's Foreign Policy Involvement

The international breakthrough of Greenland came with the Second World War, when Greenland became underneath US protection. The first Greenlandic involvement was an agreement with the USA regarding the defense of the island in 1941. This agreement was made possible by the Danish ambassador in Washington D.C. at the time, Henrik Kauffmann (Ackrén and Jakobsen, 2015; Boel and Thuesen, 2010). The defense agreement regulated the security policy between the USA and Greenland with the USA taking a hold in Greenland by establishing military bases and controlling the waters around Greenland. The governors in Greenland were taking the supreme authority over Greenland at the time, since the Danish Government was not able carrying out its task due to German occupation (Boel and Thuesen, 2010; Heinrich, 2018). The agreement on the defense of Greenland was renewed in 1951. The Americans began to establish the Thule Air Base (now Pituffik Space Base) in the northernmost part of Greenland during the beginning of the 1950s (Ackrén, 2019). The defense agreement was covering the whole of North Atlantic and not just Greenland, but Greenland is mentioned several times in the document (see *Defense of Greenland: Agreement between the United States and the Kingdom of Denmark, April 27, 1951*).

During the 1950s and 1960s a decolonization period started due to the UN system of dissolving the former colony structure. The former colonies could now choose to either go for some form of self-government or outright independence. In the Greenlandic context, Denmark revised its constitution in 1953 and integrated Greenland as a county amongst other counties within its jurisdiction. There was never a referendum in Greenland on this matter (Kleist, 2019). Since the 1970s Greenland has been active within the UN framework regarding issues concerning indigenous rights and was one of the founding fathers in relation to the Inuit Circumpolar Council (ICC) established in 1977.

In the referendum of membership to the EEC back in 1972, Greenland voted against accession with a 70% majority, but since Greenland was only a county Denmark forced Greenland into the EEC (Gad, 2013). After the referendum, an urge for more autonomy was on the agenda and in 1979 Home Rule was established. The campaign against the EEC continued and in 1982 a new referendum was held where 53% of the majority voted in favor of leaving the EEC. The negotiations started between European, Danish, and Greenlandic officials and three years later

DOI: 10.4324/9781003387657-4

a "Greenland Treaty" was in place and the island became underneath the OCT[1] system (Gad, 2013). Greenland is not an EU-member, but Greenland has access to various EU programs as an OCT-member and has also signed bilateral agreements with the EU. The OCT-status has existed since the establishment of the EC in 1957. This category was originally referred to as remnants of European imperialism, but nowadays constitute territories which have received some form of self-governance (Gad et al., 2011:15–16; Hannibal et al., 2013:77).

With the introduction of Home Rule in 1979 some powers were transferred to the Greenlandic Government and within the foreign policy affairs Greenland was to be heard before Denmark was signing any treaty or international agreement where Greenlandic affairs might be affected (Jensen and Heinrich, 2017:389). This can be said to be a notification principle, but the decisions were still made in Copenhagen.

In 1984, Greenland became a member of the Nordic Council. The Government is appointing two members, which are members of the Council and at the Nordic Council of Ministers, the ministers take part depending on the issue on the agenda. Greenland has also taken the initiative for establishing the West Nordic Council, a parliamentarian cooperation between the Faroe Islands, Greenland, and Iceland in 1985 (Jensen and Heinrich, 2017:393). The West Nordic Council can only give recommendations to the respective parliaments. However, some of these recommendations might become implemented in the regions and debated in their respective parliaments.

Since 1994, Greenland has had a foreign office underneath the Home Rule. The increasing foreign policy interest from the Greenlandic side was leading to an upgrade of the foreign office to become a foreign policy directorate in 2004 and later in 2018 this became what is today the Ministry of Foreign Affairs (Kleist, 2019).

A committee at the official level, the so-called Kleist/Friis-committee was established in 1995 to outline the cooperation between Greenland and Denmark regarding foreign- and security policy. This committee is a forum for informing Greenland what is going on in the foreign- and security matters that might affect Greenland (*Udenrigspolitisk redegørelse 2004*). This agreement is a Danish/Greenlandic agreement, and it is stated that there should be at least two meetings every year within this forum.

In May 2002, a working group was established underneath the Kleist/Friis-committee to go through the Defense Agreement from 1951. After a meeting in Washington D.C. between the US Foreign Minister, the Danish Foreign Minister, and the Greenlandic counterpart it was stated that the USA wanted an upgrade of the agreement. The Greenlandic Parliament debated about this matter in April 2003 and the result was to give permission to the USA for an upgrade of the Thule Air Base. The negotiations led to the Itilleq Declaration (*Udenrigspolitisk redegørelse 2004*). The Itilleq Declaration was signed in 2003 between the then Danish Foreign Minister, Per Stig Møller, and the Greenlandic Premier, Hans Enoksen, and relates to matters regarding Greenlandic influence within foreign policy. The declaration states that there should be a factual voice and influence from the Greenlandic side as an equal partner (Jensen and Heinrich, 2017:391). The Igaliku Agreement of

2004 amends and supplements the 1951 treaty between the USA, Denmark, and Greenland and in this case include Greenland as a cosignatory of the agreement (Ackrén, 2019). The Joint Committee is a result of the Igaliku Agreement and includes not only the upgrade of the Thule Air Base, but also cooperation regarding environmental issues, and technical and economic cooperation (Ackrén, 2014). The difference between the Itilleq Declaration and the Igaliku Agreement is that the former is a national, bilateral agreement between Denmark and Greenland, while the latter is an international, trilateral agreement between the USA, Denmark, and Greenland.

In 2005, the Danish Parliament instituted an authorization act for both Greenland and the Faroe Islands. This piece of legislation gave the Greenlandic Parliament power to negotiate and sign treaties within the areas of the Home Rule Act without Danish influence (Jensen and Heinrich, 2017). The Act has been used to sign bilateral treaties between several countries, especially within fishery and trade. Greenland has bilateral treaties with countries, such as, Iceland, Faroe Islands, Russia,[2] China, and Japan.

From 2006 to 2014 enhanced attention towards the Arctic and High North became on the agenda. This can be illustrated by the reaction to the Russian demonstrative planting of a titanium flag on the North Pole seabed in 2007 (Rahbek-Clemmensen, 2018). Another aspect is the development of Arctic strategies from all Arctic states and non-Arctic states (Heininen, 2012). The increase of attention is also related to the high market prices on natural resources that were on its hype during 2012–2013. We can say that there has been a shift in the past two decades regarding Arctic security. A lot of different internal and external images, visions and perceptions have been used to illustrate the Arctic in form of an “Eldorado”, a “sink for pollutants”, a “tipping point for the earth system”, a “race for resources”, a “global Arctic”, or a “homeland”. Other narratives also prevail such as “high latitudes-low tension”, “emerging conflicts”, “great power rivalry”, “zone of peace” and “Arctic exceptionalism” or a “new Cold War” (Heininen et al., 2019).

The Danish strategy has been reflecting the Arctic as a window of opportunity for continuous provision of services for the Greenlandic population and having its military presence in the High North. This has further been emphasized in the Danish Arctic strategies and defense agreements. The relationship towards the USA is also a vital pillar of Denmark’s Arctic presence. The Danish Armed Forces in the Arctic include missions regarding sovereignty enforcement, patrolling, surveillance, environmental protection, fisheries inspection, and search and rescue. One of the recent campaigns by the army is also to recruit Greenlanders into the Armed Forces’ activities (Rahbek-Clemmensen, 2014). The Greenlandic authorities are not able to handle all potential risks or natural disasters arising from climate change, increased commercial activities both on land and on sea or other possible contingencies. Therefore, cooperation is done in the field of SAR and civil preparedness between the Danish Armed Forces, Government of Greenland, the police and other departments.

Today, the Ministry of Statehood and Foreign Affairs manages the overarching coordination, contact and advice within the Government’s foreign affairs relations. The matters cover Arctic cooperation, the EU, Nordic cooperation, indigenous peoples, bilateral cooperation, foreign- and security policy as well as territorial and maritime matters (incl. maritime law). The Ministry is giving advice about international relations to various departments' resort areas if they require advice. This can be about working practices, different international fora, international law or concrete cases. The Ministry as already mentioned is working closely with the Ministry of Foreign Affairs in Denmark especially when Greenlandic matters are on the table. The cooperation should be according to full participation from the Greenlandic side in such matters. Furthermore, the Ministry has the responsibility to develop, coordinate and uphold the Government's interests in foreign policy. The Ministry also cooperates closely with the international representations abroad and the representation in Copenhagen is servicing the diplomatic corps abroad. There are today about 20 persons working within the Ministry of Foreign Affairs including the representatives of the "mini embassies" abroad. Internship programs are also available in Reykjavik, Brussels, Washington D.C. and Nuuk for Greenlandic students (*Udenrigspolitisk redegørelse 2022*:14). In this way, the Ministry tries to build up a new generation of diplomats for future endeavors. It can be said that the Ministry is one of the few ministries that has been able to build up a Greenlandic diplomatic corps, since all representatives abroad are Greenlandic and therefore serve Greenlandic interests with a deep knowledge of Greenlandic culture, traditions and politics.

## 4.1 Greenland's Ministers and Premiers Over Time

Greenland follows a multiparty system where no parties receive a majority of the votes, hence coalition governments are therefore established. The winning party must negotiate with the other parties to form a majority government. In Greenland we regularly see shifting coalitions during the election periods.[3] This means that we experience a lot of exchanges within the government coalitions and ministers' portfolios. This is illustrated by Table 4.1, where only the governments, Premiers, coalitions and the Minister of Foreign Affairs (MFA: s) occur. This also means that the structure and organization of the Departments or Ministries can change over time. For example, the organization of the Ministry of Foreign Affairs changed during April 2017, so the Department of Foreign Affairs became part of the Ministry of Statehood, Foreign Affairs and Agriculture. A new minister was also appointed for this latter department, Suka K. Frederiksen (*Udenrigspolitisk redegørelse 2017*:9–12). In 2018, a new coalition was formed after the election on 24th of April 2018. The election resulted in changes of the ministers and again the Ministry of Foreign Affairs became an own Ministry (*Udenrigspolitisk redegørelse* 2018:9).

According to the table there have been ten different MFA: s over the 20-year period. This means that on average we have an MFA for every second year. However, the table also indicates that the same person has been sitting with the portfolio several times with some exemptions in between.[4] The table also illustrates

*Table 4.1* Greenland's Premiers and Ministers of Foreign Affairs (MFA: s) over Time

| ***Period*** | ***Premier*** | ***Coalition*** | ***MFA*** |
|---|---|---|---|
| 20 January–13 September 2003 | Hans Enoksen | Siumut, Atassut | Hans Enoksen |
| 13 September 2003–1 December 2005 | Hans Enoksen | Siumut, Inuit Ataqatigiit (IA) | Josef Motzfeldt |
| 1 December 2005–1 May 2007 | Hans Enoksen | Siumut, IA, Atassut | Josef Motzfeldt |
| 1 May 2007–12 June 2009 | Hans Enoksen | Siumut, Atassut | Hans Enoksen (2008)<br>Kuupik Kleist (2009) |
| 12 June 2009–5 April 2013 | Kuupik Kleist | IA, Demokraatit, Kattusseqatigiit Partiiat (Alliance of Candidates) | Kuupik Kleist |
| 5 April 2013–5 November 2013 | Aleqa Hammond | Siumut, Atassut, Partii Inuit | Kuupik Kleist |
| 5 November 2013–1 October 2014 | Aleqa Hammond | Siumut, Atassut | Aleqa Hammond |
| 1 October 2014–12 December 2014 | Kim Kielsen | Siumut | Aleqa Hammond |
| 12 December 2014–27 October 2016 | Kim Kielsen | Siumut, Atassut, Demokraatit | Vittus Qujaukitsoq |
| 27 October 2016–15 May 2018 | Kim Kielsen | Siumut, IA, Partii Naleraq | Suka K. Frederiksen (2017)<br>Vivian Motzfeldt (2018) |
| 15 May 2018–5 October 2018 | Kim Kielsen | Siumut, Partii Naleraq, Nunatta Qitornai, Atassut | Vivian Motzfeldt |
| 10 April 2019–29 May 2020 | Kim Kielsen | Siumut, Nunatta Qitornai | Ane Lone Bagger (2019)<br>Steen Lynge (2020) |
| 29 May 2020–23 April 2021 | Kim Kielsen | Siumut, Demokraatit (to 8 February 2021), Nunatta Qitornai | Pele Broberg (2021) |
| 23 April 2021–4 April 2022 | Múte B. Egede | IA, Naleraq | Vivian Motzfeldt (2022) |
| 5 April 2022–present | Múte B. Egede | IA, Siumut | Vivian Motzfeldt (2022–present) |

*Source:* Maria Ackrén and Uffe Jakobsen (2024: 151); *Udenrigspolitisk redegørelser 2003–2023.*

that the Premier has had the portfolio of MFA in some periods. Since the coalition governments tend to change between elections, it can be hard to maintain a clear direction within the foreign- and security policy. However, as we will see from the annual foreign policy papers (Chapter 5), a general view of development seems to exist and that is to evolve into more independent decision-making within the foreign- and security policy sphere. Even though some MFA: s might have been stating their own policies at times, the overall direction seems to be common. Therefore, the direction Greenland has taken is quite straightforward into more independence within the foreign- and security policy.

## 4.2 Greenland's Relationship with Denmark

Greenland's relationship with Denmark is outlined both in the Danish Constitution and the Self-Government Act from 2009. The Danish Constitution holds that it is the Kingdom of Denmark, who is responsible for the overall foreign- and security policy of the whole Kingdom (including the Faroe Islands and Greenland). However, there are some grey zones, which will be outlined later in the text. In the Home Rule Act back in 1979–2009, it was stated quite clearly that if the Home Rule Government was to decide and struck agreements with foreign powers or participate in international cooperation, this was to be negotiated with the central authorities before acceptance of such an arrangement (Hviid, 2022:54). There was in this sense a clearance mechanism between the Danish and Greenlandic authorities at that time. In 2005 as earlier mentioned a new piece of legislation was introduced as the Authorization Act, which gave both the Faroe Islands and Greenland the possibility to negotiate international treaties in relation to those competence areas that had been transferred to both regional governments (Lov nr. 577 af 24. juni 2005). This legislation has been interpreted by the Faroe Islands and Greenland as the possibility to sign bilateral agreements with foreign powers in areas of vital importance for both regions. Bilateral agreements have been struck in areas of fisheries and trade with several countries as mentioned above.

The Self-Government Act of 2009 gives Greenland room for maneuvering in international relations according to Chapter 4 of the Act. "Naalakkersuisut may, on behalf of the Realm, negotiate and conclude agreements under international law with foreign states and international organizations, including administrative agreements which exclusively concern Greenland and entirely relate to fields of responsibility taken over" (Art. 12:1 Act on Greenland Self-Government, Act no. 473 of 12 June 2009). Naalakkersuisut should inform Denmark whenever an international negotiation with a foreign power is taken place and when an agreement is reached. This information goes both ways. The Danish Government should also inform Greenland whenever an international agreement is of vital importance for Greenland that is underway (Art. 12:5 and art. 13:1 Act on Greenland Self-Government, Act no. 473 of 12 June 2009). Furthermore, Greenland can negotiate membership in international organizations where not only states are members. The Act on Greenland Self-Government seems to have incorporated parts of the Authorization Act in its Chapter 4.

The Danish Constitution stipulates that it is the Kingdom of Denmark, who controls foreign- and security policy, citizenship, finance- and currency policy, and the supreme court. These areas cannot be overtaken by Greenland. The situation between the Danish Constitution and the Self-Government Act has been debated amongst officials, politicians and academics to be interpreted in two major ways: one according to a law perspective where only the delegation principle is focused on, which means that the Danish government could withdraw the Self-Government Act if it chooses to do so, but the other way is to see the Self-Government Act as a symbol of a treaty where the political and societal contexts are taken into account and where both Denmark and Greenland are treated as "equal partners". This latter interpretation has become the one which most people support and see as most realistic. The treaty perspective can therefore be seen as a federal principle (Justinussen, 2019).

Regarding the formulations of foreign policy issues within the Realm (Kingdom of Denmark) we see a differentiation of statements in relation to the bilateral agreements. The following statements are seen in the documents (Table 4.2):

This means, of course, that sometimes misunderstandings can occur between the relationships with foreign powers, where states are used to handling international agreements with only central authorities and not subnational regions. Greenland and the Faroe Islands act on their own regarding their bilateral agreements without influence from Denmark.

With the Ilulissat Declaration from 2008 Denmark became more aware of its position as an Arctic state. The Danish Arctic strategy from 2011 states the Danish interests in the Arctic as an active and serious actor on Arctic issues (Jacobsen, 2021), however, without Greenland, Denmark would not be seen as a major Arctic state. Denmark has intensified its own work towards the Arctic through various dimensions. The Danish Arctic strategy from 2011 was the first Arctic strategy by the Kingdom of Denmark and it also contained a merger of previous Faroe Islands Command and Greenland Command (established in 1961 and 1951 respectively). On 31 October 2012 this became realized as a Joint Arctic Command established in Nuuk (Strandsbjerg and Dahlberg, 2024:49). However, this initiative was merely a cut in defense spending in the Arctic than a strategic matter. The military budget was cut from 2.1% to 1.3% of GDP between 1998 and 2011 (Strandsbjerg and Dahlberg, 2024:54). The so-called Taksøe-Jensen investigation in 2016 gave some proposals for how Denmark could give more priorities towards the foreign- and security political interests in the Arctic. A short time thereafter a new Danish government saw Arctic as one of the main top priorities within foreign- and security matters. This led to that a double of employees became to work with the Arctic within the Ministry of Foreign Affairs in Denmark to replace some co-workers

*Table 4.2* The Kingdom of Denmark's Formulations within Foreign Policy in Relation to Bilateral Agreements

| ***Legal formulation*** | ***Covers*** |
|---|---|
| Kingdom of Denmark | International agreements that are valid for the whole Realm (including Denmark, the Faroe Islands and Greenland) |
| Kingdom of Denmark in the case of the Faroe Islands | International agreements, which only cover the Faroe Islands |
| Kingdom of Denmark in the case of Greenland | International agreements, which only cover Greenland |
| Kingdom of Denmark in the case of the Faroe Islands and Greenland | International agreements, which only cover the Faroe Islands and Greenland (but not Denmark) |

*Source*: Rasmus Gjedsø Bertelsen and Jens Christian Svabo Justinussen (2020), 'Færøernes international politik', table 1, *Økonomi & Politik*, Årg. 93, nr. 4/2020 p. 40 (author's own translation from the Danish text).

from the Asian part. More funding was allocated to the Arctic Command and more events were focused on the Arctic as a region (including the ten-year celebration of the Ilulissat Declaration in 2018), and new Arctic financial mechanisms became introduced. Moreover, funding was given yearly between 2019 and 2022 to implement the Arctic Hub in Nuuk as a research hub (Jacobsen, 2021). These initiatives have also spilled over to Greenland in the form of being more active in foreign- and security matters. Today we see that the Kingdom of Denmark is seeking to enhance the military and international defense cooperation in the Arctic through the cooperation with the Joint Arctic Command, but also within the Arctic Coast Guard Forum and other forms of cooperation through SAR.[4]

The recent Arctic strategy from the Government of Greenland publicly outlined in March 2024 states that Greenland should be the main actor for instance within the Arctic Council and Greenland should work to keep the Arctic as a low-tension area with a focus on establishing even a Center for Peace. Other main points in the strategy are that democratic rules and human rights should be kept in all relations and that Greenland should develop the living conditions for its people. All relationships should follow the principle of "nothing about us without us" (*Grønland i verden – intet om os uden os*, 2024). Copenhagen has been waiting for the Greenlandic strategy before publishing the Kingdom of Denmark's overall Arctic strategy for the whole Realm (including the Faroe Islands and Greenland). The new Danish strategy is still underway.

### *4.2.1 Grey Zones in Foreign- and Security Policy*

The extent to which Greenland's and Denmark's interests coincide depends on the case and therefore some grey zones can appear. One such example was the "Uranium debate", which had its hype in 2010s. In the 2000s, the discourse of 'zero-tolerance' in Greenland was coined. However, there has never been any official document stating this matter. In 2009 with the introduction of the new Self-Government Act, Greenland received competence over all its natural resources. A new Mineral Resources Act was implemented because of this in 2010. The uranium mining also has international implications, since the extraction of uranium is underneath international conventions and IAEA[5] monitoring, which Denmark has responsibility over. In April 2013, 48 NGOs from around the world signed an appeal to the Greenlandic and Danish governments to keep the zero-tolerance policy in place (Bjørst, 2017). However, on Thursday 24 October 2013, a decision made by the Greenlandic Parliament was a narrow favor (by 15–14) of repealing the zero-tolerance policy on the mining of uranium and other radioactive materials (Nuttall, 2013). This came as a surprise for the Danish Government. Denmark sees the extraction as part of the foreign- and security dimension, while Greenland looks at it as part of internal, domestic policy in relation to other natural resources.

As mentioned by Bjørst (2016) there are two major discourses in Greenlandic politics when it comes to uranium mining. The risk analysis or discourse concentrated around the issues of how uranium mining will destroy the local community and global nature as such. This discourse is usually framed by NGOs and hesitant local people. The other discourse is in line with the Government of Greenland to look for a brighter future with economical gains from the mining industry and where the local community will be saved with new job opportunities and benefit the Greenlandic economy as such (Bjørst, 2016). These discourses are very much in conflict with each other.

The issue, however, came to an end in 2016, when the Government of Greenland and the Government of Denmark signed an agreement regarding the uranium issue. This was based on an earlier report, which had been launched in 2013 regarding the issue of entering an agreement complex between Denmark and Greenland of future cooperation in relation to exploitation and export of uranium from Greenland (*Udenrigspolitisk redegørelse 2017*:74–75). This can be seen as a compromise and gives Greenland the right to exploit and extract uranium, while Denmark will stand responsible for upholding the international conventions and regulations within the field of extracting radioactive material. There is overlapping legislation in this field, since we do have laws which are underneath the Greenlandic Government (such as the Mineral Resources Act), the Self-Government Act, the Danish constitution, Danish law, and added to this international law regarding nuclear non-proliferation, some of which applies to Denmark but not to Greenland. In practice, Greenlandic authorities will issue licenses for pre-studies, investigations and exploitation of uranium and other radioactive material. Any company or other actors involved in these activities will deal with the Greenlandic authorities only. However, nuclear safety is a common competency of both governments. The self-government has the responsibility for all nuclear facilities that are built and run in connection with mining and the direct processing of uranium (*Aftale med den danske regering om uran*). The Danish foreign ministry and the Greenlandic self-government will cooperate closely and share information both ways on all issues relating to the nexus of nuclear material and foreign-, security-, and defense policy (*Aftale med den danske regering om uran*). There has not been a mine opening yet in relation to extracting uranium, but the potential is there.

Another grey zone has occurred in relation to the nexus of the relationships between the USA, Denmark and Greenland. The USA has been pressuring Denmark to hinder China entering Greenland several times. China was interested in buying 'Grønnedal' a former navy base in the south of Greenland in 2016, and there were also interests from Chinese entrepreneurs in investing in the new airport in Nuuk, which also was objected by the USA in 2018. Denmark has therefore invested financial support for the civilian airport in Nuuk instead (Jakobsen, 2021:134). This has however intervened with domestic interests in a Greenlandic context and therefore some disputes have occurred between Nuuk and Copenhagen regarding these issues. The trust between the Government of Greenland and the Danish Government has therefore been compromised at times. It has not always been clear from a Greenlandic standpoint that these matters de facto included some security concerns in relation to the USA and Denmark. The case of leaving the former naval base was totally underneath the decision of the Danish military authorities and the airport was seen as a security matter from the US side (Olsvig, 2022a:223).

"The Arctic capacity package" is DKK 1.5 billion for enhancing the security in the Arctic, which was decided upon in 2021 by the Danish Parliament (*Folketinget*). This meant that the capacities both in Greenland and the Faroe Islands was to be enhanced by 2023 (Kjærgaard, 2021:219). The "Arctic capacity package" is financing a reopening of the radar at the Faroe Islands, *Sornfelli*, and this gives the Danish Defense valuable surveillance of the air area between the Greenlandic east coast over Iceland and the Faroe Islands to Scotland. The financial package is also enhancing the capacity for maritime surveillance of Greenland (Kjærgaard, 2021:228). The Danish defense has several tasks in the Arctic, i.e. assertion of sovereignty, search and rescue services, maritime environmental tasks, and support for the civil society (Nielsen, 2021:242).

### *4.2.2 Greenland's Boycott from the Arctic Council in 2013*

In the Arctic Council, Greenland has been part of the Danish delegation on an equal footing with Denmark. However, at the Kiruna meeting 2013 Sweden as having the chairmanship was deciding that only states should be sitting around the table. Before that Denmark had three seats around the table with representatives from Denmark, Faroe Islands and Greenland respectively with their own flags as a symbol. The Kiruna meeting led to a boycott from Greenland with the then Premier of Greenland, Aleqa Hammond from Siumut and this led to suspension of all Greenland's activities within the Arctic Council, since this was seen as a degraded role of Greenland as an Arctic nation. After three months with the following Canadian chairmanship negotiations led to an agreement, which gave the opportunities for both Greenland and the Faroe Islands to join the table again as far as possible (Jacobsen, 2021). A compromise was reached, where if there are fewer seats than three for Kingdom of Denmark, the representative, who has the most vital issues on the table should take the lead. Nowadays, it is only the Danish flag on the table (Jacobsen, 2019; Jacobsen, 2021).

## 4.3 Greenland's Relationship with the USA

Looking back in history Greenland has been mentioned by official US geostrategic interest already in 1832, when President Andrew Jackson's administration had the idea to buy Greenland. Three decades later, Secretary of State William H. Seward rearticulated the idea when he commissioned a report about the resources in Greenland and Iceland (Jacobsen and Olsvig, 2024:110). In 1917, when the Danish West Indies were to be transferred from Denmark to the USA, Greenland was again on the table, but Denmark was stating demands that Greenland should be underneath the sovereignty of Denmark. Denmark was to have the control over the island, but the demands was not totally acknowledged by the USA and rumors flourished about the USA trying to buy the island once again. The Danish Prime Minister, Thorvald Stauning, found it necessary to issue a statement in *Financial Times* that "Greenland was not for sale" in 1930 (Jacobsen and Olsvig, 2024:118). When the World War II broke out in 1939, high-ranking US decision-makers once again discussed the issue of buying Greenland and when Denmark became occupied by Nazi-Germany in 1940, the new era began with Greenland underneath the protection of the USA. However, Greenland was never sold to the Americans due to the diplomatic relations and the agreement that came into place between the Danish ambassador Henrik Kaufmann and the USA regarding the defense and security policy in 1941 (Jacobsen and Olsvig, 2024:119). However, the 1941 agreement provided the US military with nearly unlimited rights. Greenland had become an area where the US military basically could do whatever they wanted without any effective parliamentary control (Heidbrink, 2022:19–20). The USA saw Greenland as an essential region for the American defense and hereby the Monroe Doctrine was activated in relation to Greenland. With the end of WWII, Denmark requested a withdrawal of American troops from Greenland and an annulment of the 1941 agreement, but the Truman administration gave Denmark a counterproposal, where three alternatives were laid out:

> 1) a continuation of the 1941 agreement, where U.S. was supposed to take over the whole defense of the island, 2) a lease of existing U.S. bases in Greenland for the next 99 years or 3) a purchase of Greenland for the price of US$100 million in gold.

The Danish Government rejected all three proposals, but in 1948 the Danish Government gave up and gave the USA the opportunity to continue with its military presence in Greenland and this led to a new defense agreement in 1951 (Jacobsen and Olsvig, 2024:121–122; Heidbrink, 2022:21).

Since the beginning of the Cold War, the USA has had three geopolitical interests directed towards Greenland: (1) detection of cross-polar attacks by strategic missiles and bombers, (2) control over the Greenland-Iceland-United Kingdom (GIUK) gap, and (3) access to Greenlandic minerals. However, the latter has only been of interest lately because of Chinese investments within several mining projects going on in Greenland (Rahbek-Clemmensen, 2020:177). The Pituffik Space Base (former Thule Air Base) is US military's most northern installation and of importance for the ballistic missile defense system. The GIUK gap, which covers the waters surrounding Greenland, Iceland, the Faroe Islands, and the British Isles, are a point where NATO can detect and block hostile fleets' passage into the North Atlantic. It is often seen as a blind spot where Russian submarines might occur. The assets of rare earth elements (REEs) located in South Greenland are seen as critical resources for the US Department of Energy (Rahbek-Clemmensen, 2020:177–178).

Agreements with Washington D.C. and the American presence in Greenland, generally outlines the military activities that occur within and around Greenland. The "Permanent Committee", which was established in 1991 is where the USA, Greenland and Denmark are discussing these matters to exchange information and have a constructive dialogue about the US military presence (Olsvig, 2022a:222).

In recent years the USA has had several initiatives directed towards the Arctic. The USA has reestablished the base in Keflavik in Iceland, which was closed back in 2006. The Americans have also reestablished their second fleet in the

North Atlantic, especially in surveilling the area around the GIUK gap (Rabehk-Clemmensen and Sørensen, 2021:19). Furthermore, the USA has invested a help package to Greenland of DKK 83 million in 2020. The USA has also a new service agreement in relation to Pituffik Space Base and a partnership agreement with Greenland (Jakobsen, 2021:137). The Americans have also reopened a US Consulate in Nuuk since June 2020.

Copenhagen has had to accommodate the great powers of the USA and others to secure its sovereignty over Greenland. Danish sovereignty over the island was fully settled after the Permanent Court of International Justice in the Hague ruled against a Norwegian claim to eastern Greenland back in 1933. The "Greenland question" has been one of the reasons why Denmark became a founding member of NATO in 1949 (Rahbek-Clemmensen, 2020:178–179).

The current relationship between Greenland and the USA is based on the following documents:

- The Defense Agreement of 1951
- The so-called Igaliku Agreement from 2004 (incl. Joint Declaration on Economic and Technical Cooperation and Joint Declaration on Environment)
- The Kangerlussuaq Agreement
- The Dundas Agreement
- Memorandum of Understanding of 2006
- A Memorandum of Understanding from 2019 regarding the Mineral Resource Sector

#### *4.3.1 The Defense Agreement of 1951*

The Defense Agreement from 1951 regulates the relationship between the Government of Denmark and the Government of the USA regarding Greenland as an area of military and security interest. This agreement relates to the American bases in Greenland at the time. The USA had built several military bases around Greenland: Grønnedal south of Nuuk, which later became underneath the Danish armed forces; Kangerlussuaq in the middle of the country, which is used for both civilian and military purposes nowadays; Narsarsuaq in South Greenland, which nowadays is used as a civilian airport; two bases were also placed in East Greenland, which have been left and destroyed; and then the Pituffik/Thule Air Base in North Greenland, which is the only base still in operation today. The Defense Agreement can be interpreted to only regulate the activities in Pituffik/Thule Air Base currently and for future endeavors.

Article 1 in the Defense Agreement states that the Government of Denmark and the Government of the USA will promote stability and well-being in the North Atlantic Treaty region by coordinating collective forces in order to secure peace and security and take necessary steps if an attack occur in Greenland in accordance with NATO plans (*Defense of Greenland: Agreement between the United States and the Kingdom of Denmark, April 27, 1951*). The Agreement can therefore also be seen as a NATO agreement as it covers Greenland, but also other NATO regions in the Arctic (such as the Faroe Islands and Iceland).

#### *4.3.2 The Igaliku Agreement of 2004*

The Igaliku Agreement signed August 6, 2004, is an amended and supplemented version of the previous Defense Agreement of 1951. According to Article 1 it is clearly stated that Thule Air Base is the only defense area in Greenland. The flag of the Kingdom of Denmark, the flag of Greenland and the flag of the USA should fly over Thule Air Base (*The Igaliku Agreement 2004*). The three partners in this agreement are: the Kingdom of Denmark, Greenland, and the USA. All parties agree that the NATO Status of Forces Agreement (NATO SOFA) has been valid since 1955 in Greenland and that this agreement should be seen as a supplementary agreement between the parties involved. The USA should consult with and inform the Government of Denmark and the Government of Greenland prior to any important changes made in military operations or military facilities in Greenland that are underneath American rule (*The Igaliku Agreement 2004*).

The Igaliku Agreement is escorted by a joint declaration regarding economic and technical cooperation and a joint declaration on cooperation within environment in Greenland. The joint declaration on economic and technical cooperation considers the ties between Greenland and the USA from a historical point of view and in this view hopes to strengthen the relationships between the two nations. Cooperation in areas such as research, technology and energy, environmental issues, education, development, tourism, air traffic planning and trade are mentioned as possible areas of enhanced cooperation. It is stated that a Joint Committee should be established to promote and coordinate the wide spectrum of areas of cooperation. Joint projects and programs will be implemented through mutual contributions. The Joint Committee with representatives from all three partner countries (Denmark, Greenland, and the USA) should meet annually (*The Igaliku Agreement 2004*). The joint declaration on cooperation on environment in Greenland considers the prevention and combatting of pollution dangerous to human health and other pollution of air, ice, water, and earth. It also considers the protection of nature, its wild animals, plants, and their inhabitants, as well as protection of the landscape and areas of historical and scientific value. The US armed forces should also respect the Greenlandic environmental standards regarding the Thule area (*The Igaliku Agreement 2004)*.

#### *4.3.3 The Kangerlussuaq Agreement*

The so-called Kangerlussuaq Agreement is a Memorandum of Understanding between the Government of the Kingdom of Denmark (including the Home Rule Government of Greenland) and the Government of the USA regarding the use of Kangerlussuaq aviation facilities, Kulusuk airfield and other matters related to the US military activities in Greenland (*The Kangerlussuaq Agreement*). This Memorandum of Understanding was signed in 1991. This MoU provides for a transition of responsibility from US forces to Danish/Greenlandic authorities regarding the operation and support of Kangerlussuaq and Kulusuk. After the transition phase the US forces will continuously have the right to use the facilities at Kangerlussuaq and Kulusuk for occasional missions, exercises or in contingency (*The Kangerlussuaq Agreement*).

The US military authorities will have responsibility over all US military aircraft, other US property, and US forces personnel. In case of accidents occurring to US military aircraft in Greenland, US military authorities and Danish authorities will cooperate in the adoption of rescue missions (*The Kangerlussuaq Agreement*).

#### *4.3.4 The Dundas Agreement*

The Dundas Agreement is an agreement between the Department of Environment at the Danish Government and the Greenlandic Government regarding the environmental mitigating efforts being done at the Dundas area. This agreement is not between the US government, but only a domestic agreement between Denmark and Greenland. However, the USA and the Thule Air Base might be affected by the activities taken place in Dundas. There is a need for some cooperation with the Pituffik/Thule Air Base and the use of its facilities (*The Dundas Agreement*). The Dundas Agreement is part of the cleaning up of waste and pollution surrounding the area. The document was signed in 2011. Usually, the polluter pays principle is used in such circumstances, but the reluctance of the USA taking its responsibility for clean-up, this agreement was made only between Denmark and Greenland.

#### *4.3.5 Memorandum of Understanding of 2006*

The MoU from 2006 considers cooperation within scientific and technical research regarding earth sciences. This MoU has been struck between the US Geological Survey (USGS) of the Department of the Interior of the USA, the Geological Survey of Denmark and Greenland (GEUS) of the Ministry of the Environment of Denmark, and the former Bureau of Minerals and Petroleum (BMP) of the Home Rule of Greenland. The purpose of the MoU is to provide a framework for the exchange of information and scientific and technical knowledge between USGS, GEUS, and BMP (*The Memorandum of Understanding 2006*). The MoU is an addition to the previously mentioned joint declarations related to the Igaliku Agreement. A MoU is not a legally binding agreement. It is more of an expression of the will two or more parties might have and can be seen more as a formal form of a gentlemen's agreement.

#### *4.3.6 The Memorandum of Understanding from 2019 Regarding Mineral Resource Sector*

The MoU establishes cooperation between the Greenland's Ministry of Mineral Resources and Labor and the US Department of State. This MoU is focused on increased cooperation within the field of mineral resources in the form of implementing sound resource sector development and governance, promote investment and advance the deployment of new technologies. The participants (Greenland and the USA) will try to mobilize their respective expertise to assist in sound mineral developments in Greenland and work together to foster cooperation between different stakeholders from both the public and private sphere (*The MoU 2019*).

#### *4.3.7 Recent Developments*

The enhanced interest towards the Arctic has been seen as a shift within US policy from 2010. Before 2010, the USA used to be represented by Under-Secretaries or Deputy Secretaries of State within Arctic forums, but with Secretary of State Hillary Clinton we see a shift where the USA begin to prioritize high-level Arctic forums. Hillary Clinton participated in the Arctic Five meeting in Chelsea, Canada in 2010 and then in the Arctic Council Ministerial Meeting in Nuuk, Greenland in 2011. John Kerry continued this trend by participating in the Arctic Council ministerial meetings in 2013 and 2015, and he visited Greenland in 2016. President Obama's visit to Alaska in 2015 also shows an interest towards Arctic matters. The Obama administration sought to put climate change on the political agenda in Washington D.C. and Greenland was then used as the reference point of the severe implications of the climate change and the melting of the icecaps (Henriksen and Rahbek-Clemmensen, 2017).

The recent developments regarding the relationship between Greenland and the USA has been focused on the issue of former US President Trump's idea of purchasing Greenland, which came about in August 2019. It is not the first time the USA has been trying to buy Greenland, a previous attempt happened in 1800s and 1900s as previously mentioned. However, the idea was rejected by the Greenlandic Government with the slogan: "We are open for business, but not for sale" and the Danish counterpart, where the Prime Minister Mette Frederiksen stated that the idea was "absurd" (Olsvig and Nielsen, 2019).

The interest in having closer relationships with Greenland from the US side has been realized in other ways. A re-opened US consulate is now in place in Nuuk, the University has received a US corner, and economic support for technical and mineral activities has been approved. The enhanced interest towards the Arctic is for and foremost a reaction towards other great powers, such as, Russia and China as potential competitors over Greenlandic assets of for example rare earth elements (REEs). Another issue is to enhance the relationships directly with Greenlandic authorities in an event of potential Greenlandic independence. This collaboration is stated in a "Common Plan" struck between the Government of Greenland and the Government of the USA, which was finalized in 2020 (Olsvig, 2022b:102). One of the objectives is to have enhanced cooperation within bilateral trade and investment. The two governments will work together to resolve any existing regulatory and tariff barriers to increase Greenlandic market access for goods and businesses. Another aim is to enhance commercial partnerships between the USA and Greenlandic businesses across diverse sectors, especially within tourism, seafood and fisheries, and natural resource extraction. A third aim is to leverage a US University to work with counterparts in Greenland to build capacity in academic programs and vocational training in land and fisheries management, hospitality, and sustainable tourism (*Common Plan for U.S.-Greenland Cooperation*).

The shift with the Biden administration has confirmed that the USA is not interested to buy Greenland anymore. The Biden administration has recently been suggesting a new environmental review of potential oil and gas development in the Arctic National Wildlife Refuge in Alaska after Trump's administration review, which had many flaws in it. The plateau in Alaska has been off limits to oil development for decades. This changed in 2017 when the Republican majority in the Congress approved an oil and gas leasing program. Now, the Biden administration will restore the ban of oil and gas development in the area (Fountain, 2021). It seems that the Biden administration is now trying to restore the relationships with their allies in all parts of the world, where previous President Trump had spoiled the contacts or made the USA as an unreliable partner.

The Joint Committee, which is the platform for structured government-to-government cooperation between the USA and Greenland, had a meeting on September 28, 2023, in Qaqortoq in South Greenland. Here both partners reaffirmed their commitment to enhance economic ties; furthering educational and cultural exchange and scientific cooperation; and coordinating and collaborating closely in the energy and mineral resource sectors for the benefit of both peoples. The meeting was a review of meeting specific objectives laid out in the "Common Plan" from 2020 (*U.S.-Greenland Joint Committee Statement, September 28, 2023*).

As the close and direct dialogue between Greenland and the USA has increased, we see that Greenland is using its own maneuvering space and provides the USA with a position to engage directly with Greenland without Danish influence. With the new elected Trump administration since November 2024, we will see what will happen. The Government of Greenland has stated that cooperation with the USA will continue, but it is of course uncertain in which form this will evolve for the next four years.

## 4.4 Greenland's Relationship with the EU

As already mentioned in the beginning of the chapter Greenland was a member of the EEC back in 1973–1979 due to the status as a county within the Danish Kingdom. Even though Greenland had voted no for a membership in 1972, Greenland became a member of the EEC together with Denmark. With the implementation of the Home Rule in 1979 a new referendum was held in 1982 and thereafter Greenland was leaving the EEC. It took three years of negotiations before an agreement was reached. Greenland became an OCT within the EU framework. Greenland has succeeded in having favorable agreements with the EU, especially within fisheries and education.

Greenland's overall relationship with the EU consist of a package of various protocols and agreements:

1. A protocol in relation to the EU Treaty regarding the secession from EU
2. An association agreement regarding the OCT-status, hereunder access to EU programs
3. A fishery agreement with a fishery protocol
4. A political common declaration
5. A partnership agreement with focus on education
6. Participation in the EU: s Northern Periphery Program

7 Free custom agreement regarding specific fishery products
8 Export of Greenlandic raw diamonds underneath the Kimberley process
9 Veterinary protocol
10 Letter of intent regarding the natural resources (*Udenrigspolitisk redegørelse 2013*:13).

The above-mentioned protocols and agreements get renewed before a period ends apart from the first protocol, which dates to 1985. The representation in Brussels opened in 1992. This was the first mission abroad from Greenland. The representation has the following agenda:

- To secure visibility in Brussels and in cooperation with the Greenlandic Government and to cooperate closely with the EU Parliament, EU Commission and the Council
- Be the contact between the self-government and the EU members/ambassadors in Brussels and relevant lobbyists
- Coordinate issues between Greenland and the EU institutions. In practical terms this is outlined as daily contacts with the Danish EU representation, where issues are coordinated which have relevance for Greenland
- The representation is used as a learning platform for young students, who have an interest for Greenlandic international relations (*Udenrigspolitisk redegørelse 2014*:18).

### *4.4.1 "The Arctic Window" Initiative 1999*

The "Arctic Window" initiative in 1999 was an initiative taken by the then sitting Premier of Greenland, Jonathan Motzfeldt. This was in the aftermath of the Faroese election of 1998, where the Republican Party took the victory, and the Faroe Islands were to work for independence as a goal. The idea behind the "Arctic Window" initiative was to enhance the engagement of the EU towards the North, especially in Arctic matters. This was also in line with the Finnish idea that had occurred in 1997 with the "Northern Dimension". The Finnish idea was however a tactic to receive more attention towards cooperation between the northern regions and Russia (Hviid, 2022:45, 50). From a Greenlandic standpoint the aim with the "Arctic Window" was to receive an "Arctic desk officer" in the European Commission. The idea behind this was to build up the knowledge and expertise about the Arctic in Brussels. However, this initiative was never realized. The timing was not the best from an EU standpoint, since the EU was working with the extension towards the East with new countries coming in as member states (Hviid, 2022:53). The focus on the Arctic was minimal at this point.

Today, the situation is different. The EU has its own Arctic strategy and has recently (2024) opened an EU Commission office in Nuuk as part of the strategy. The office will manage EU support to Greenland and facilitate deeper partnership with Greenland and the European Commission in close cooperation with the

Special Envoy for Arctic Matters (European Commission, 2021:4). Within the EU Arctic strategy, the EU is emphasizing to contribute to maintain a peaceful and constructive dialogue and cooperation to keep Arctic safe and stable. The focus is on addressing the ecological, social, economic and political challenges arising from climate change and in this way the EU is actively involved in tackling climate change and environmental degradation, making the Arctic more resilient through environmental legislation and goals. Furthermore, it is stated that the EU is willing to support the sustainable development of the Arctic region for the benefit of the inhabitants and future generations, focusing on the needs of the Indigenous Peoples (European Commission, 2021:2). In the 2021–2027 program, Greenland will receive €225 million per year. The EU also has a fisheries agreement with Greenland giving its fleets access to Greenlandic waters (European Parliament, 2022). Economically, the EU is very important for Greenland, since the support coming from the EU is the largest amount after the Danish block grant (about DKK 3.6 billion) as foreign direct investment.

## 4.5 Greenland's Involvement in International Organizations

Greenland is involved in many different international organizations and institutions, especially in the field of fishery and marine management, but also in relation to indigenous matters through the Inuit Circumpolar Council and the UN framework. Other important organizations are the Arctic Council, the Nordic Council and the Nordic Council of Ministers and the West Nordic Council.

### *4.5.1 The Arctic Council*

The initial effort for a widespread Arctic cooperation was launched in 1987 in Murmansk by the Soviet Union's then-Secretary-General Mikhail Gorbachev. His famous speech about the Arctic as a peace zone also included statements about cooperation regarding the protection of the Arctic environment (Koivurova and Vanderzwaag, 2007).

In 1989, Finland put Arctic cooperation on its agenda by inviting officials from the eight Arctic states to Rovaniemi, Lapland to discuss measures to protect the Arctic environment. The efforts that followed to establish a more comprehensive structure for Arctic cooperation are sometimes referred to as the Rovaniemi Process (Pedersen, 2012). By 1991 an agreement had been reached on an *Arctic Environmental Protection Strategy* (AEPS) (Arnold, 2012; Koivurova, 2010; Pedersen, 2012). Even though, the AEPS is often seen as a Finnish initiative, Canadian officials had a major role in the development of the founding documents of the AEPS as this was linked to the Canadian domestic policy of the Arctic Environmental Strategy (Arnold, 2012). The Canadian strategy was developed through cooperation with Canadian indigenous peoples of the north.

The Inuit Circumpolar Conference (now Inuit Circumpolar Council), Nordic Saami Council (now Saami Council) and the USSR Association of Small Peoples of the North (now Russian Association of Indigenous Peoples of the North,

RAIPON) were accorded observer status in the AEPS. Germany, Poland, the UK, the UNECE, the United Nations Environment Program (UNEP), and the International Arctic Science Committee (IASC) were also added as observers in the AEPS (Fenge, 2013).

The observer status in the AEPS was not given permission to attend meetings of SAAOs (Senior Arctic Affairs Officials, nowadays SAOs), where ministers were discussing recommendations. At the first ministerial meeting of the AEPS in Nuuk, Greenland 1993, the political leaders of the Indigenous organizations breakfasted privately with ministers, where they expressed their dissatisfaction of being excluded from the cooperation in the Arctic affairs. The ministers agreed and afterwards these organizations were permitted to attend and intervene in all meetings of the AEPS, and this was the start for the 'Permanent Participants' – a term first formally used in the AEPS Inuvik Declaration of March 1996 (Fenge, 2013; Koivurova, 2010). The meeting in Nuuk led to the Nuuk Declaration of Environment and Development in the Arctic and was signed by the Arctic-eight. The working groups' mandates were extended, and a new Task Force on Sustainable Development and Utilization (TFSDU) was established. The Danish government promised to create a secretariat to enhance the participation of the representatives of the indigenous peoples and this became eventually the Arctic Council's Indigenous Peoples' Secretariat (IPS) (Koivurova and Vanderzwaag, 2007). The IPS was first placed in Copenhagen, Denmark, but with the permanent secretariat of the Arctic Council now in operation in Tromsø, Norway, IPS has also moved to the same location, so everything is gathered at one and the same place.

The AEPS can be considered as the forerunner for the Arctic Council. The AEPS had five main objectives:

> to protect the Arctic ecosystem including humans; to provide for the protection, enhancement and restoration of environmental quality and the sustainable utilization of natural resources, including their use by local populations and indigenous peoples in the Arctic; to recognize and, to the extent possible, seek to accommodate the traditional and cultural needs, values and practices of the indigenous peoples as determined by themselves, related to the protection of the Arctic environment; to review regularly the state of the Arctic environment; and to identify, reduce, and as a final goal, eliminate pollution.
> (Bloom, 1999)

As part of the AEPS the Arctic states established four working groups: the Conservation of Arctic Flora and Fauna (CAFF); the Emergency Prevention, Preparedness and Response (EPPR); the Arctic Monitoring and Assessment Program (AMAP); and the Protection of the Arctic Marine Environment (PAME) (Bloom, 1999; Koivurova, 2010; Pedersen, 2012). A new working group was established because of the Arctic Council realization in 1996; the Sustainable Development Working Group (SDWG) (Koivurova, 2010). The final AEPS ministerial meeting took place in Alta, Norway after the Arctic Council was already established. All the working groups have been integrated into the Arctic Council from the AEPS. Recently, some critics have been raised regarding the validation of the working groups if they are the right ones in a time of climate change and globalization (Exner-Pirot et al., 2019).

The first Arctic Council meeting took place in Iqaluit, Canada in September 1998. Here the Arctic Council Rules of Procedure and Arctic Council Terms of Reference for a Sustainable Development Program were adopted (Koivurova and Vanderzwaag, 2007; www.arctic-council.org). The Aleut International Association (AIA) was accepted as a Permanent Participant at this meeting and in 2000 the Gwich'in Council International (GCI) and Arctic Athabaskan Council (AAC) followed (Fenge, 2013). It was agreed that Permanent Participants would sit around the same table as member states, including ministerial meetings. These organizations would also be able to intervene and speak according to the same procedures as governing members (Fenge, 2013). The Arctic Council is dependent on the SAOs (Senior Arctic Officials), since they are the ones preparing ministerial meetings and coordinate the work in the Council (Koivurova, 2010). The SAOs are also leading the coordination of the working groups. The Arctic Council plays a huge role in producing various materials, such as press releases, official statements, declarations, reports and keeps its website up to date (Dodds, 2012). The council has played an important role in developing and disseminating various scientific assessments (Young, 2010). In a way the Arctic Council does function from a bottom-up perspective through its working groups doing scientific assessments and coming up with recommendations for the ministerial level. The Arctic Council bases its decision on the principle of consensus amongst its members and for now there have been challenges to operate accordingly, because of the Russian invasion of Ukraine. However, no one is excluded from the Arctic Council, but there has been some pause in the work (Retter, 2024:28, 31).

The current Norwegian chairmanship is soon coming to an end in May 2025. Norway has been able to reestablish some of the work within the Arctic Council through written procedures and online working group meetings. Kingdom of Denmark will take over as the next chair for the Arctic Council and Greenland will therefore play a prominent role as the only part of the Kingdom being situated in the Arctic.

### *4.5.2 The Inuit Circumpolar Council*

In the Arctic, the indigenous mobilization started in the 1970s, when comprehensive land claims and self-government agreements or home rule arrangements became implemented in most of the regions inhabited by indigenous peoples in the North (for example, Nunavik, Alaska, Nunavut and Greenland). Furthermore, the rush for oil, gas and minerals during the 1960s and 1970s in the Arctic made it clear that changes were to happen for the Arctic peoples (Dahl et al., 2024:8). The first Arctic Peoples' Conference was taking place in Copenhagen in 1973 by Indigenous Peoples' organizations and representatives, the University of Copenhagen and IWGIA (International Working Group for Indigenous Affairs). A total number of 40 delegates and 21 organizations took part in the event from Arctic Canada,

Greenland and Sápmi (Finland, Norway and Sweden). The vision of the conference was to demand states for recognition of political rights of Arctic Indigenous Peoples, as well as land rights, human rights, and equality (Dahl et al., 2024:5). A proposal from the conference back in 1973 was also to establish "a Circumpolar Body of Indigenous Peoples" to pursue the shared and collective interests and to be able to discuss common challenges and possibilities for the future (Dahl et al., 2024:11).

Following, this conference and parallel developments of global Indigenous Peoples' movement own organizations and representative institutions have been established. In 1977, Inuit Circumpolar Conference (today's Inuit Circumpolar Council (ICC)) was established, and the resolution was stating that the Arctic region should be seen as a peaceful and safe zone (Olsvig and Cullen, 2024b:47). The rights of Indigenous Peoples have been affirmed in the ILO Convention 169 from 1989 as well as the United Nations Declaration on the Rights of Indigenous Peoples (UNDRIP) adopted in 2007 (Dahl et al., 2024:7). Indigenous peoples are commonly considering international law as a very important tool for achieving their political goals. The reason behind that is that many countries give limited opportunities for groups to take part in the political development or political participation (Koivurova and Heinämäki, 2006). Governance amongst indigenous peoples has become used as a mechanism to refer to principles, institutions, and practices that a collective group enjoys regulating relations among its members, and between its members and the outside world. This is based on the "beliefs, values, practices, and worldviews of indigenous cultures" (Fondahl and Irlbacher-Fox, 2009:1–2). The most prominent elements of governance within the indigenous context are: "knowing through experience, merit as a source of authority, decision-making by consensus, being part of the environment, using resources wisely for the common good, and respecting diversity" (Fondahl and Irlbacher-Fox, 2009:4).

In 1970, the UN Sub-Commission on Prevention of Discrimination and Protection of Minorities recommended a study on the specific issue of discrimination against Indigenous Peoples. The Special Rapporteur at that time, José Martinez Cobo submitted several reports during 1981 to 1984 on the issue on Indigenous Peoples. This led to the establishment of the Working Group on Indigenous Populations under the Economic and Social Council within the UN framework in 1982 (Olsvig, 2024:13). Since then, many other efforts within the UN framework have been taken, for example in 2020 an establishment of the Indigenous Coordinating Body (ICB) was realized. Both the Sámi Council and the Inuit Circumpolar Council have two seats representing the Arctic region (Olsvig, 2024:17). The ICC contributes substantively to various fora related to the Arctic. As a Permanent Participant in the Arctic Council, ICC is also contributing to the various working groups underneath the Arctic Council. The ICC holds a distinct status in the Intergovernmental Panel on Climate Change (IPCC) as well as the International Maritime Organization (IMO). Furthermore, the ICC is an UN-credited NGO (Olsvig and Cullen, 2024b:43).

The ICC has initiated processes to establish Inuit-led marine governance regimes between Greenland and Canada in relation to polynya in the North Waters (*Pikialasorsuaq*). This has been formalized in a Letter of Intent between the governments of Greenland and Canada in October 2023 (Olsvig and Cullen, 2024b:43).

The fundamental aims of the ICC are to be a united voice of Inuit and to promote Inuit rights and interests at the international and national level. Furthermore, the ICC supports and ensures effective participation and meaningful engagement in political, economic, and social institutions. The work within ICC is to ensure the sustainability of Inuit culture and societies for present and future generations, and to promote long-term management and protection of Arctic and sub-Arctic wildlife, environment, biological diversity and Inuit ways of life. The ICC also works for wise management of Arctic non-renewable resources (*Inuit Circumpolar Council Charter*, Article 3). Now Greenland is having the chairship of the ICC until 2026.

### *4.5.3 Nordic Cooperation*

Greenland became a member of the Nordic Council and the Nordic Council of Ministers in 1984. There had been discussions already during the 1970s about proposals to join the Nordic family. Greenland has been very active in Nordic cooperation and had this as one of the top priorities of the Department of Foreign Affairs in the past (Creutz and Spiliopoulou Åkermark, 2021:50). The Government of Greenland presents an overview of Nordic cooperation and initiatives in various reports, such as, *Nordisk redegørelse*, the annual Foreign Policy Reports and *Årsberetning for Inatsisartut* (Creutz and Spiliopoulou Åkermark, 2021:50). As the other Nordic autonomous regions (Faroe Islands and Åland) Greenland is also represented by two appointed parliamentarians in the Nordic Council. These are considered as the voice outwards, who seek to ensure Greenlandic interests and goals within the Nordic cooperation (Creutz and Spiliopoulou Åkermark, 2021:51). Over the years, the Greenlandic delegates have participated in meetings, committees, and Council sessions. One issue, which has been focused on has been the language used within Nordic cooperation. For example, in 2016 the then Head of Delegation of Greenland, Iddimanngiiu Bianco submitted a proposal to the Nordic Council for accepting Greenlandic as an official language that could be used in negotiations and meetings with the Nordic Council and the Nordic Council of Ministers (Creutz and Spiliopoulou Åkermark, 2021:52). For now, Danish, Norwegian, and Swedish are the working languages in the Nordic Council.

According to the Declaration of Nordic Language Policy of 2007, which is not a binding document, all Nordic languages should be equal. However, they do not play the same role. A language may be *complete and essential to society* in relation to the language community where it is spoken. Greenlandic is here seen as an *essential* language for the society but is not totally *complete* in every context of society (Declaration of Nordic Language Policy, 2007), since Danish is widespread within official authorities and education. For example, all legislation and official documents are translated into both Danish and Greenlandic. Greenlandic should be supported and essential to the society also for future purposes according to the Declaration (Declaration of Nordic Language Policy, 2007).

In 2012, a more intensified cooperation between the Faroe Islands, Greenland and the Åland Islands was laid down in a memorandum of understanding (Creutz and Spiliopoulou Åkermark, 2021:53). This was to enhance the cooperation within Nordic cooperation between the autonomous regions. One major issue, which has been debated especially amongst the representatives from the Faroe Islands and Greenland, is the role of the autonomous regions as full members of the Nordic Council and the Nordic Council of Ministers. It is a question that has been flourishing every now and then, but the Nordic states have not agreed on this proposal (Creutz and Spiliopoulou Åkermark, 2021:57). Recent debates show that Greenland has now taken time out from the Nordic cooperation to give more understanding that the autonomous regions have developing their own international relations and therefore should be able to act as full members with voting rights within the Nordic Council. Finland will take over the chairmanship after Iceland in 2025 and there are some hopes that this issue should be solved either before or under the next period. Originally, it was Kingdom of Denmark, who should have taken over the chairship, but since Denmark will be chairing the Arctic Council as well as the EU, Finland has taken over the turn of the Nordic Council. This also shows some solidarity between the Nordic countries, when countries do have a lot of issues on their plate.

#### *4.5.4 West Nordic Council*

Greenland took the initiative to establish the West Nordic Council, first named as West Nordic Parliamentarian Council of Cooperation, which was implemented in 1985. This is an interparliamentary cooperation between Greenland, the Faroe Islands and Iceland. The idea is to promote ideas and recommendations for West Nordic interests. Each parliament is represented by six members of parliament (Creutz and Spiliopoulou Åkermark, 2021:54; Nielsson, 2014:3).

In 1997 a new charter was agreed upon by the three respective West Nordic national parliaments, and the Council's name was changed from the West Nordic Parliamentarian Council of Cooperation to the West Nordic Council (WNC). Subsequently, the WNC introduced new working rules and took other steps to strengthen its impact (Nielsson, 2014:7). The Council's general work and activities are administrated by a Presidium consisting of one member from each delegation – including a president, 1st vice president and 2nd vice president. The secretariat is in Iceland. The WNC convenes twice every year, once for an annual general meeting that constitutes the Council's supreme authority, and once for a thematic conference on a particular topic of shared interest (Nielsson, 2014:7; West Nordic Council). The results are expressed through recommendations approved at the annual general meeting, which are then presented to the member countries' parliaments and acted upon by the appropriate ministers (West Nordic Council; Nielsson, 2014:7). For Greenland it provides one of the few venues where it can engage in international cooperation without Denmark's supervision, another notable exception of this kind being NAMMCO (the North Atlantic Marine Mammal Commission) (Nielsson, 2014:8).

#### *4.5.5 The Involvement within the UN*

During the late 1960s waves of public protests towards racial segregation, war and imperial rule were on the international stage. In 1970, the UN Sub-Commission on Prevention of Discrimination and Protection of Minorities initiated a comprehensive study over the specific issue of discrimination against Indigenous Peoples. This work led to the establishment of the Working Group on Indigenous Populations under the auspices of the Economic and Social Council in 1982 and by 1985 preparation for a UN Declaration on the Rights of Indigenous Peoples (UNDRIP) (Olsvig and Cullen, 2024a:156). An Arctic People's Conference in 1973 was a starting point to gather Arctic Indigenous Peoples together to discuss their common interests and challenges. The Inuit Circumpolar Council (ICC) was founded four years after the conference in 1977, with the Russian Inuit joining in 1992. The Saami Council also joined the same year in 1992. The ICC sought and obtained consultative status in the UN and has actively participated in the UN processes ever since (Olsvig and Cullen, 2024a:157).

The UN Permanent Forum on Indigenous Issues was established in 2000. The UN Human Rights Council founded the Expert Mechanism on the Rights of Indigenous Peoples in the same year that the UNDRIP was adopted, and the first UN Special Rapporteur on the Rights of Indigenous Peoples was appointed in 2001. Another platform is the Local Communities and Indigenous Peoples Platform, which was established in 2017 under the auspices of the UN Framework Convention of Climate Change (Olsvig and Cullen, 2024a:158). The first World Conference on Indigenous Peoples was held at UN Headquarters in New York in September 2014 as a high-level plenary meeting of the UN General Assembly (Olsvig and Cullen, 2024a:158).

Naalakkersuisut (the Government of Greenland) is following the work in the UN and is trying to oversee that Greenland lives up to the UN Conventions and Declarations where Greenland is part. Greenland is actively involved in the Danish delegation at UN General Assemblies. Furthermore, Greenland is taken an active stand within representative organs, for example underneath the Human Rights Council and underlying committees in relation to reports and periodic examinations in relation to states' compliance with international conventions (*Udenrigspolitisk redegørelse, 2010*:20).

#### *4.5.6 International Fisheries Organizations*

Greenland is active in several international fisheries organizations, since fishery is the utmost important sector of the economy. The fishing industry provides over 90% of Greenland´s exports. The most important species are codfish, Greenland halibut, mackerel, crab and prawn. Greenland's business sector is dominated by large publicly owned companies. The largest companies with 100% ownership by the Self-Government are Royal Greenland A/S (fishing industry), KNI A/S (retail and oil sales), Royal Arctic Line A/S (shipping), Air Greenland (air traffic), and Tusass (telecommunications). About 1/3 of the turnover created by companies comes from fishing and related industry and trade. The other major industry is wholesale and retail trade, which also accounts for nearly 1/3 of the total turnover in companies (Greenland in Figures 2024).

Fishing for prawns and Greenlandic halibut and some other species is regulated by quota and license regulations decided by the government. Fishing comes in two versions: coastal – and offshore fishing. Coastal fishing supplies land-based factories, while the offshore fishing fleet primarily consists of factory vessels with on-board production (Greenland in Figures 2024).

Greenland has bilateral or trilateral fisheries agreements with the Faroe Islands, Norway and Iceland, and has a general fisheries agreement with the EU. Greenland has the legal authority regarding the fisheries sector, and Greenland's Fishery Control authority (GFLK) often deploys observers on Greenlandic and international vessels, primarily on shrimp trawlers (Greenland in Figures 2024).

Greenlandic shrimps stand for ¼ of the total supply of cold-water shrimps in Japan. Royal Greenland has had over 30 years a sales office in Tokyo. The Greenlandic export of fish and shellfish in 2020 was around DKK 30 million (US$5.7 million) to Japan. In China the amount of Greenlandic fishery products lies at DKK 1.5 billion (US$213 million) (*Udenrigspolitisk redegørelse 2022*:41–42). In 2021, 51.6% of the total export value of cooked and pilled shrimps was going to the UK. Due to Brexit, Greenland is trying to negotiate a free trade agreement with the UK (*Udenrigspolitisk redegørelse 2022*:56).

### *4.5.7 NAFO*

NAFO is the North Atlantic Fishery Organization and is an intergovernmental organization, which consists of 13 members, or so-called contracting partners. These are: Canada, Cuba, Denmark on behalf of the Faroe Islands and Greenland, EU, France on behalf of St. Pierre and Miquelon, Iceland, Japan, Norway, the UK, South Korea, Ukraine, and the USA. The organization has an area which covers the Northwest of the Atlantic, hereunder the ocean between Greenland and Canada. NAFO is regulating the oceans outside the 200 nautical miles of Greenland's southwest coast, Canada's east coast and the USA's northeast coast. The management of these areas covers all fish- and shellfish species, except for salmon. NAFO is working to secure the species and to optimize exploitation of the fishing resources. NAFO is also giving biological advice for the whole area, hereunder the Greenlandic fish- and shellfish stocks in West Greenland (*Udenrigspolitisk redegørelser 2003–2022*).

### *4.5.8 NEAFC*

NEAFC is the Northeast Atlantic Fisheries Commission and is also an intergovernmental organization, which can be seen as a sister organization for NAFO. NEAFC covers the Northeast of the Atlantic. NEAFC has six members: Denmark on behalf of the Faroe Islands and Greenland, EU, Iceland, Norway, Russia, and the UK. Greenland has traditionally had most interest in the pelagic redfish in the Irminger Sea. Greenland is also fishing mackerel, herring, and blue whiting (*Udenrigspolitisk redegørelser 2003–2022*).

#### *4.5.9 NASCO*

NASCO is an international fishery organization with seven members: Canada, Denmark on behalf of the Faroe Islands and Greenland, EU, Norway, Russia, the UK, and the USA. NASCO is standing for North Atlantic Salmon Conservation Organization and has three regional commissions: the Northeast Atlantic Commission (with EU, Denmark on behalf on the Faroe Islands and Greenland, Norway, the UK, and Russia); the North American Commission (with Canada and the USA); the West Greenlandic Commission (with Canada, EU, the UK, the USA, and Denmark on behalf of Greenland). The purpose of NASCO is to promote and secure the reestablishment and increase of salmon stocks, and to enhance scientific analyses and information of the stocks. Greenland is working to secure its historical rights for the fishery of salmon. It is forbidden to export salmon from Greenland (*Udenrigspolitisk redegørelser 2003–2022*).

#### *4.5.10 Icelandic-Greenlandic Fishery Commission*

Greenland and Iceland have since January 2013 signed an agreement, which has established the Icelandic-Greenlandic Fishery Commission. Iceland and Greenland share several fish stocks and therefore it has been important to cooperate to keep a sustainable level of fishing stocks in the waters around Iceland and Greenland. Furthermore, Iceland and Greenland are discussing matters in relation to NAFO, NEAFC and other forums of interest within the commission (*Udenrigspolitisk redegørelser 2013–2023*).

#### *4.5.11 NA-FIG*

NA-FIG is the North Atlantic Fishery Intelligence Group, which became established in 2015. This organization works in close cooperation with the OECD and the Nordic Council of Ministers to prevent and combat the increasing organized crimes within the fishery sector. NA-FIG has been able to create necessary attention to the needs of established networks between intelligence services between authorities both nationally and internationally. Greenland has acceded the protocol in February 2020 and has been established a working group cross departments in the Government to handle these issues (*Udenrigspolitisk redegørelser 2020–2022*).

#### *4.5.12 International Fishery Agreement in the Arctic High Ocean*

Ten large fishery nations have signed an international agreement in October 2018, which should prevent unregulated fishery operations in the Central Arctic Ocean. The signatories are Canada, the USA, Russia, Norway, Denmark on behalf of the Faroe Islands and Greenland, Iceland, EU, South Korea, Japan, and China. The agreement regulates fisheries in the Arctic High Ocean. The partners of the agreement commit themselves to not fishing in the area. The agreement is valid for 16 years and thereafter the agreement will be renewed automatically each 5th year, if none of the partners reject such an incident. Underneath the agreement a Joint Program of scientific Research and Monitoring in the Arctic High Sea will be established (*Udenrigspolitisk redegørelser 2018–2022*).

#### *4.5.13 International Mammal Organizations*

Greenland has a long hunting tradition and therefore Greenland is also active in international organizations related to large mammals in the world. Hunting has been a way of life in Greenland for generations. Even today, hunting provides an important supplement to the economy of the families. Hunting is regulated by means of seasons and permissions. A hunting license is mandatory for anyone who wants to hunt. The general license comes in two forms: professional and recreational. In addition, a specific license is needed when hunting species, which are limited by quota. The quota system regulates the number of animals available for hunting and is decided by the Government of Greenland (Greenland in Figures 2024).

#### *4.5.14 NAMMCO*

North Atlantic Marine Mammal Commission (NAMMCO) is an organization where Greenland, Norway, Iceland and the Faroe Islands are members. NAMMCO was founded in 1992 and works to secure, manage and study marine mammals regionally in the North Atlantic. Other countries, such as, Canada, Denmark, Russia and Japan participate as observers in NAMMCO. The right to hunt and exploit marine mammals is the cornerstone of NAMMCO. The principles of hunting and exploiting animals should follow sustainable principles so that the animals are not suffering, and that hunting equipment is used in a sensible manner. Hunting methods and improvements of these have a central place in the discussions within NAMMCO (*Udenrigspolitisk redegørelser 2003–2023*).

#### *4.5.15 IWC*

The International Whaling Commission (IWC) has been operating since 1946 and is a global forum to secure the right form of management of the World's largest whales. In 1986 it was decided to forbid commercial exploitation of whale products. Greenland is a member through Denmark's membership. The main purpose is to provide for the proper conservation of whale stocks and thus make possible the orderly development of the whaling industry (https://iwc.int/commission/history-and-purpose).

#### *4.5.16 Joint Commission on Narwhal and Beluga*

The Joint Commission on Narwhal and Beluga (JCNB) is between Canada and Greenland from 1989. This rests on a MoU between Canada/Nunavut and Greenland with NAMMCO as an observer. JCNB gives biological and managerial advice about the stocks of narwhal and beluga in Baffin Bay. NAMMCO gives managerial advice in David Strait and East Greenland. Research is conducted in Greenlandic waters (*Udenrigspolitisk redegørelser 2003–2023*).

#### *4.5.17 Oslo Convention on Polar Bears*

The 1973 Convention on the Conservation of Polar Bears was signed by Denmark/Greenland, Canada/Nunavut, Norway, Russia, and the USA/Alaska in May 1976. Greenland gives reports on the quotas of polar bears and has included local knowledge in hunting and exploitation of polar bears. Scientific research is also going on through Greenland Institute of Natural Resources (Udenrigspolitisk redegørelse 2021:74–75).

## Notes

1 OCT stands for Overseas Country and Territory.
2 The agreement with Russia is suspended after the Russian invasion to Ukraine as earlier mentioned.
3 For more information about the political system, see Ackrén and Jakobsen (2024:135–176).
4 SAR stands for search and rescue missions.
5 IAEA stands for the International Atomic Energy Agency.

## References

Ackrén, Maria (2014). 'Greenlandic Paradiplomatic Relations'. In Lassi Heininen (ed.): *Security and Sovereignty in the North Atlantic*. Basingstoke and New York: Palgrave Macmillan, pp. 42–61.

Ackrén, Maria (2019). 'From Bilateral to Trilateral Agreement: The Case of Thule Air Base'. In Lassi Heininen, Heather Exner-Pirot, and Justin Barnes (eds.): *Redefining Arctic Security: Arctic Yearbook*. Akureyri: Arctic Portal, pp. 33–42. Available at: https://www.arcticyearbook.com

Ackrén, Maria and Uffe Jakobsen (2015). 'Greenland as a Self-governing Sub-national Territory in International Relations: Past, Current and Future Perspectives', *Polar Record*, 15:4, pp. 404–412.

Ackrén, Maria and Uffe Jakobsen (2024). 'The Capacity of Self-Government in Greenland'. In Jan Sundberg and Stefan Sjöblom (eds.): *Governing Partially Independent Nation-Territories: Evidences from Northern Europe*. Cham: Palgrave Macmillan, pp. 135–176.

Act on Greenland Self-Government (2009). Act no. 473 of 12 June 2009. Available at: https://www.stm.dk/media/11338/10-notifikation-af-7-oktober-2009-til-fn-s-generalsekretaer-om-selvstyreloven.pdf

*Aftale med den danske regering om uran*. Available at: https://www.retsinformation.dk/eli/ft/201512L00155

Arctic Council. Available at: https://arctic-council.org/

Arnold, Samantha (2012). 'Constructing an Indigenous Nordicity: The "New Partnership" and Canada's Northern Agenda', *International Studies Perspectives* 13:1, pp. 105–120.

Bertelsen, Rasmus G. and Jens Christian Svabo Justinussen (2020). 'Færøernes internationale politik', *Økonomi & Politik*, Årg. 93, nr. 4/2020, pp. 37–48.

Bjørst, Lill R. (2016). 'Saving or Destroying the Local Community? Conflicting Spatial Storylines in the Greenlandic Debate on Uranium', *The Extractive Industries and Society*, 3:1, pp. 34–40.

Bjørst, Lill R. (2017). 'Uranium: The Road to "Economic Self-Sustainability for Greenland"? Changing Uranium-Positions in Greenlandic Politics'. In Gail Fondahl and Gary N. Wilson (eds.): *Northern Sustainabilities: Understanding and Addressing Change in the Circumpolar World*, Springer Polar Sciences. Cham: Springer International Publishing, pp. 25–34.

Bloom Evan, T. (1999). 'Establishment of the Arctic Council', *The American Journal of International Law*, 93:3, pp. 712–722. https://www.jstor.org/stable/2555272 (Accessed 23 February 2016).

Boel, Jens and Søren T. Thuesen (2010). 'Greenland and the World: The Impact of World War II on Danish-Greenlandic Relations' (1993). In Karen Langgård, Flemming Nielsen, Birgit Kleist Pedersen, Kennet Pedersen, Jette Ryggaard (eds.): *Cultural and Social Research in Greenland - Selected Essays 1992–2010*. Nuuk: Ilisimatusarfik/Forlaget Atuagkat, pp. 9–35.

*Common Plan for U.S.-Greenland Cooperation*. Available at: https://naalakkersuisut.gl/-/media/publikationer/selvstaendighed_og_udenrigsanliggender/2020/commonplanforusgreenland.pdf (Visited 17 September 2021).

Creutz, Katja and Sia Spiliopoulou Åkermark (2021). *The Faroe Islands, Greenland and the* Åland *Islands in Nordic Cooperation*. FIIA Report December 2021/69. Helsinki: FIIA and the *Åland* Islands Peace Institute.

Dahl, Jens, Áslat Holmerg, Sara Olsvig, and Kathrin Wessendorf (2024). 'Introduction'. In Jens Dahl, Áslat Holmberg, Sara Olsvig and Kathrin Wessendorf (eds.): *Empowering Arctic Indigenous Peoples: Celebrating 50 Years of Indigenous Diplomacy*. Copenhagen: IWGIA-Saami Council-ICC, pp. 5–11.

Declaration of Nordic Language Policy (*Deklaration om nordisk språkpolitik*) (2007). Köpenhamn: Nordiska Ministerrådet. Available at: https://norden.diva-portal.org/smash/get/diva2:700895/FULLTEXT01.pdf

*Defense of Greenland: Agreement between the United States and the Kingdom of Denmark, April 27, 1951*. Available at: https://avalon.law.yale.edu/20th_century/den001.asp

Dodds, Klaus (2012). 'Anticipating the Arctic and the Arctic Council: Pre-emption, Precaution and Preparedness'. In: T. S. Axworthy, T. Koivurova and W. Hasanat (eds.): *The Arctic Council: Its Place in the Future of Arctic Governance*. Collaboration between the Institute for Northern and Environmental Studies at the University of Lapland, Canada Centre for Global Security Studies, Munk School of Global Affairs, University of Toronto and the Walter and Duncan Gordon Foundation. E-book: Toronto: The Gordon Foundation, Munch School of Global Affairs, pp. 1–28. Available at: https://gordonfoundation.ca/resource/the-arctic-council-its-place-in-the-future-of-arctic-governance/

European Commission (2021). *A Stronger EU Engagement for a Peaceful, Sustainable and Prosperous Arctic*. Joint Communication to the European Parliament, the Council, the European Economic and Social Committee and the Committee of the Regions. High Representative of the Union for Foreign Affairs and Security Policy. Brussels. JOIN (2021) 27 Final.

European Parliament (2022). Briefing: *EU Regional Policy in the Arctic*, June 2022.

Exner-Pirot, Heather, Maria Ackrén, Natalia Loukacheva, Heather Nicol, Annika E. Nilsson, and Jennifer Spencer (2019). 'Form and Function: The Future of the Arctic Council'. Available at: https://www.thearcticinstitute.org/form-function-future-arctic-council/

Fenge, Terry (2013). 'The Arctic Council: Past, Present, and Future Prospects with Canada in the Chair from 2013 to 2015', *The Northern Review*, No. 37 (Fall 2013), pp. 7–35.

Fondahl, Gail and Stephanie Irlbacher-Fox (2009). *Indigenous Governance in the Arctic*. A Report for the Arctic Governance Project, November 2009. Toronto: The Walter and Duncan Gordon Foundation.

Fountain, Henry (2021). 'The Biden Administration Orders a New Review of Oil and Gas Development in the Arctic', *The New York Times*, August 3, 2021. Available at: https://www.nytimes.com/2021/08/03/us/politics/environmental-review-arctic-refuge.html (Visited 21 September 2021).

Gad, Ulrik P. (2013). 'Greenland Projecting Sovereignty – Denmark Protecting Sovereignty Away'. In Rebecca Adler-Nissen and Ulrik Pram Gad (eds.): *European Integration and Postcolonial Sovereignty Games. The EU Overseas Countries and Territories*. London and New York: Routledge, pp. 217–234.

Gad, Ulrik P., Ida Hannibal, Kristine Holst, and Rebecca Adler-Nissen (2011). 'EUs oversøiske lande og territorier: postkoloniale suverænitetsspil og Grønlands arktiske muligheder', *Tidsskriftet Politik*, 14:1, pp. 15–24.

*Greenland in Figures 2024*. Statistics Greenland. Available at: https://stat.gl/publ/en/GF/2024/pdf/Greenland%20in%20Figures%202024.pdf

*Grønland i verden – intet om os uden os* (2024). Available at: https://www.ft.dk/samling/20231/almdel/UPN/bilag/178/2827876.pdf

Hannibal, Ida, Kristine Holst, Ulrik Pram Gad, and Rebecca Adler-Nissen (2013). 'European Union – Facilitating the OCTs in Brussels'. In Rebecca Adler-Nissen and Ulrik Pram Gad (eds.): *European Integration and Postcolonial Sovereignty Games*. London and New York: Routledge, pp. 77–95.

Heidbrink, Ingo (2022). '"No One Thinks of Greenland": US-Greenland Relations and Perceptions of Greenland in the US from the Early Modern Period to the 20th Century', *American Studies in Scandinavia*, 54:2 (December 2022), pp. 8–34. https://doi.org/10.22439/asca.v54i2.6739

Heininen, Lassi (2012). *Arctic Strategies and Policies: Inventory and Comparative Study*. Akureyri: The Northern Research Forum & University of Lapland. Available at: https://www.rha.is/static/files/NRF/Publications/arctic_strategies_7th_draft_new_20120428.pdf

Heininen, Lassi, Heather Exner-Pirot, and Justin Barnes (2019). 'Redefining Arctic Security: Military, Environmental, Human or Societal? Cooperation and Conflict?'. In Lassi Heininen, Heather Exner-Pirot, and Justin Barnes (eds.): *Redefining Arctic Security: Arctic Yearbook*. Akureyri: Arctic Portal, pp. 7–14. Available at: https://www.arcticyearbook.com

Heinrich, Jens (2018). 'Independence through International Affairs: How Foreign Relations Shaped Greenlandic Identity before 1979'. In Kristian Søby Kristensen and Jon Rahbek-Clemmensen (eds.): *Greenland and the International Politics of a Changing Arctic – Postcolonial Paradiplomacy between High and Low Politics*. London and New York: Routledge, pp. 28–37.

Henriksen, Anders and Jon Rahbek-Clemmensen (2017). 'The Greenland Card: Prospects for and Barriers to Danish Arctic Diplomacy in Washington'. In Kristian Fischer and Hans Mouritzen (eds.): *Danish Foreign Policy Yearbook 2017*. Copenhagen: Danish Institute for International Studies (DIIS), pp. 75–98.

Hviid, Alexander Niels-Jacob (2022). 'Det arktiske vindue – et kig til Grønlands udenrigspolitik', *Politica*, 54:1, pp. 44–63.

*Inuit Circumpolar Council Charter*. Available at: https://www.inuitcircumpolar.com/icc-international/icc-charter/

Jacobsen, Marc (2019). 'Arktis: Grønlands strategiske arena for større udenrigspolitisk suverænitet', *Politica*, 51:4, pp. 485–506.

Jacobsen, Marc (2021). 'Danmarks arktiske udenrigspolitik: regionale ambitioner og rigsfællesskabets komplikationer', *Økonomi & Politik*, Årg. 94, nr. 2/2021, pp. 35–53.
Jacobsen, Marc and Sara Olsvig (2024). 'From Peary to Pompeo: The History of United States' Securitizations of Greenland'. In Marc Jacobsen, Ole Waever, and Ulrik Pram Gad (eds.): *Greenland in Arctic Security – (De)securitization Dynamics under Climate Thaw and Geopolitical Freeze*. Ann Arbor: University of Michigan Press, pp. 107–148.
Jakobsen, Peter V. (2021). 'Danmark har gode muligheder for at mindske stormagtskonkurrencen i Arktis'. In Jon Rahbek-Clemmensen og Camilla T. N. Sørensen (red.): *Sikkerhedspolitik i Arktis og Nordatlanten*. København: Djøf Forlag, pp. 127–147.
Jensen, Einar L. and Jens Heinrich (2017). 'Fra hjemmestyre til selvstyre 1979-2009'. In Hans Christian Gulløv (ed.): *Grønland: Den arktiske koloni*, 1st Edition, 2nd Issue. København: Gads Forlag A/S, pp. 372–421.
Justinussen, Jens Christian S. (2019). 'Rigsfællesskabet i et føderalt perspektiv', *Politica*, 51:4, pp. 441–468.
Kjærgaard, Steen (2021). 'Dansk militær opbygning i Grønland og fastholdelse af arktis lavspænding'. In Jon Rahbek-Clemmensen og Camilla T. N. Sørensen (red.): *Sikkerhedspolitik i Arktis og Nordatlanten*. København: Djøf Forlag, pp. 219–237.
Kleist, Mininnguaq (2019). 'Grønlands udenrigspolitik og internationale relationer: Nuværende rammer og mulig udvikling i et selvstændighedsperspektiv', *Politik*, 22:1, pp. 84–101.
Koivurova, Timo (2010). 'Limits and Possibilities of the Arctic Council in a Rapidly Changing Scene of Arctic Governance', *Polar Record*, 46:2, pp. 146–156.
Koivurova, Timo and Leena Heinämäki (2006). 'The Participation of Indigenous Peoples in International Norm-making in the Arctic', *Polar Record*, 42:2, pp. 101–109.
Koivurova, Timo and David L. Vanderzwaag (2007). 'The Arctic Council at 10 Years: Retrospect and Prospects', *UBC Law Review*, 40:1, pp. 121–194.
Lov nr. 577 af 24. juni 2005. Available at: https://www.retsinformation.dk/eli/lta/2005/577
Nielsen, Anders P. (2021). 'Kampen om flåden – kystvagt eller krig i farvandene omkring Grønland'. In Jon Rahbek-Clemmensen og Camilla T. N. Sørensen (eds.): *Sikkerhedspolitik i Arktis og Nordatlanten*. København: Djøf Forlag, pp. 239–261.
Nielsson, Egill T. (2014). *The West Nordic Council in the Global Arctic*. Working Papers. Institute of International Affairs. Centre for Arctic Policy Studies. Reykjavik: University of Iceland.
Nuttall, Mark (2013). 'Zero-Tolerance, Uranium and Greenland's Mining Future', *The Polar Journal*, 3:2, pp. 368–383.
Olsvig, Sara (2022a). 'Greenland's Ambiguous Action Space: Testing Internal and External Limitations between US and Danish Arctic interests', *The Polar Journal*, 12:2, pp. 215–239. https://doi.org/10.1080/2154896X.2022:2137085
Olsvig, Sara (2022b). "Odd Couples' Win-Sets Maintaining U.S. Basing Rights through New Two-Level Game Negotiatiions with Greenland', *Scandinavian Military Studies*, 7:1, pp. 93–110. https://doi.org/10.31374/sjms.208
Olsvig, Sara (2024). 'Keynote: Enhanced Participation of Indigenous Peoples in the UN'. In Jens Dahl, Áslat Holmberg, Sara Olsvig and Kathrin Wessendorf (eds.): *Empowering Arctic Indigenous Peoples: Celebrating 50 Years of Indigenous Diplomacy*. Copenhagen: IWGIA-Saami Council-ICC, pp. 13–20.
Olsvig, Sara and Miriam Cullen (2024a). 'Arctic Indigenous Peoples and International Law', *Nordic Journal of International Law*, 93, pp. 152–169.
Olsvig, Sara and Miriam Cullen (2024b). 'Arctic Indigenous Peoples and International Diplomacy: Celebrating 50 Years of Diplomatic Leadership on the International Stage'. In Jens Dahl, Áslat Holmberg, Sara Olsvig and Kathrin Wessendorf (eds.): *Empowering*

*Arctic Indigenous Peoples: Celebrating 50 Years of Indigenous Diplomacy*. Copenhagen: IWGIA-Saami Council-ICC, pp. 39–49.

Olsvig, Sara and Rasmus Leander Nielsen (2019). 'Da Trump ville købe Grønland', *Udenrigs - Tidsskrift*, 3: 74–83.

Pedersen, Torbjørn (2012). 'Debates over the Role of the Arctic Council', *Ocean Development & International Law*, 43:2, pp. 146–156.

Rahbek-Clemmensen, Jon (2014). *Arktiske Usikkerheder – Fem trusler mod det fredelige samarbejde i det høje nord*. København: DIIS Rapport 2014:20. Available at: https://www.econstor.eu/bitstream/10419/120407/1/798547111.pdf

Rahbek-Clemmensen, Jon (2018). 'The Arctic Turn: How Did the High North Become a Foreign and Security Priority for Denmark'. In Kristian Søby Kristensen and Jon Rahbek-Clemmensen (eds.): *Greenland and the International Politics of a Changing Arctic – Postcolonial Paradiplomacy between High and Low Politics*. London and New York: Routledge, pp. 54–69.

Rahbek-Clemmensen, Jon (2020). 'Denmark and Greenland's Changing Sovereignty and Security Challenges in the Arctic'. In Gunhild Hoogensen Gjørv, Marc Lanteigne, and Horatio Sam-Aggrey (eds.): *Routledge Handbook of Arctic Security*. London and New York: Routledge, pp. 176–187.

Rahbek-Clemmensen, Jon and Camilla Tenna Nørup Sørensen (2021). 'Introduktion – stormagtskonkurrencen i Arktis og Nordatlanten og dens implikationer'. In Jon Rahbek-Clemmensen og Camilla T. N. Sørensen (eds.): *Sikkerhedspolitik i Arktis og Nordatlanten*. København: Djøf Forlag, pp. 17–37.

Retter, Gunn-Britt (2024). 'Keynote: Arctic Council'. In Jens Dahl, Áslat Holmberg, Sara Olsvig and Kathrin Wessendorf (eds.): *Empowering Arctic Indigenous Peoples: Celebrating 50 Years of Indigenous Diplomacy*. Copenhagen: IWGIA – Saami Council – ICC, pp. 28–33.

Strandsbjerg, Jeppe and Rasmus Dahlberg (2024). 'Creating the Joint Arctic Command: Cutting Cost or Preparing for the Future? Strategic Culture, Arctic Security and Strategic Reasoning within the Kingdom of Denmark', *Scandinavian Military Studies*, 7:1, pp. 48–61. Available at: https://doi.org/10.31374/sjms.228

*The Dundas Agreement*. Available at: https://2009-2017.state.gov/documents/organization/165196.pdf

*The Igaliku Agreement 2004*. Available at: https://www.state.gov/wp-content/uploads/2019/02/04-806-Denmark-Defense.done_.pdf

*The Kangerlussuaq Agreement*. Available at: https://naalakkersuisut.gl/-/media/filer/formandens-departement/qallunaatut.pdf

*The Memorandum of Understanding 2006*. Available at: https://www.state.gov/wp-content/uploads/2019/02/06-531-Denmark-Sci-Coop-Greenland.done_.pdf

The MoU 2019. Available at: https://govmin.gl/wp-content/uploads/2020/04/MoU_Govt_of_Greenland_USA_2019.pdf

*Udenrigspolitisk redegørelse 2003*. FM 2003/12. Nuuk: Landsstyret/Government of Greenland.

*Udenrigspolitisk redegørelse 2004*. FM 2004/20. Nuuk: Landsstyret/Government of Greenland.

*Udenrigspolitisk redegørelse 2005*. FM 2005/29. Nuuk: Landsstyret/Government of Greenland.

*Udenrigspolitisk redegørelse 2006*. FM 2006/18. Nuuk: Landsstyret/Government of Greenland.

*Udenrigspolitisk redegørelse 2007*. FM 2007/35. Nuuk: Landsstyret/Government of Greenland.

*Udenrigspolitisk redegørelse 2008*. FM 2008/41. Nuuk: Landsstyret/Government of Greenland.
*Udenrigspolitisk redegørelse 2009*. FM 2009/14. Nuuk: Naalakkersuisut/Government of Greenland.
Udenrigspolitisk redegørelse 2010. FM 2010/14. Nuuk: Naalakkersuisut/Government of Greenland.
*Udenrigspolitisk redegørelse 2011*. FM 2011/14. Nuuk: Naalakkersuisut/Government of Greenland.
*Udenrigspolitisk redegørelse 2012*. FM 2012/14. Nuuk: Naalakkersuisut/Government of Greenland.
*Udenrigspolitisk redegørelse 2013*. EM 2013/14. Nuuk: Naalakkersuisut/Government of Greenland.
Udenrigspolitisk redegørelse 2014. EM 2014/14. Nuuk: Naalakkersuisut/Government of Greenland.
*Udenrigspolitisk redegørelse 2015*. EM 2015/14. Nuuk: Naalakkersuisut/Government of Greenland.
*Udenrigspolitisk redegørelse 2016*. EM 2016/14. Nuuk: Naalakkersuisut/Government of Greenland.
*Udenrigspolitisk redegørelse 2017*. EM 2017/14. Nuuk: Naalakkersuisut/Government of Greenland.
*Udenrigspolitisk redegørelse 2018*. EM 2018/14. Nuuk: Naalakkersuisut/Government of Greenland.
*Udenrigspolitisk redegørelse 2019*. EM 2019/14. Nuuk: Naalakkersuisut/Government of Greenland.
*Udenrigspolitisk redegørelse 2020*. EM 2020/14. Nuuk: Naalakkersuisut/Government of Greenland.
*Udenrigspolitisk redegørelse 2021*. EM 2021/14. Nuuk: Naalakkersuisut/Government of Greenland.
*Udenrigspolitisk redegørelse 2022*. EM 2022/14. Nuuk: Naalakkersuisut/Government of Greenland.
*Udenrigspolitisk redegørelse 2023*. Nuuk: Naalakkersuisut/Government of Greenland.
*U.S.-Greenland Joint Committee Statement,* September 28, 2023. Available at: https://dk.usembassy.gov/u-s-greenland-joint-committee-statement-september-28-2023/# (Visited 8 April, 2024).
*West Nordic Council*. Available at: https://www.vestnordisk.is/om-vestnordisk-rad/
Young, Oran R. (2010). 'Arctic Governance - Pathways to the Future', *Arctic Review on Law and Politics*, 1:2: 164–185.

# 5 Foreign Policy Papers from 2003 to 2023

This chapter will focus on the main empirical work regarding the foreign policy papers that the Government of Greenland has published annually between 2003 and 2023. In this way we will see the development in the recent two decades. I will divide the papers in three different time periods: the first time period will be from 2003 to 2008, which will resemble the time before Greenland had received the extended self-government that became implemented in 2009; the second time period will illustrate the time from 2009 to 2014, ending with the boycott of the Arctic Council meeting in Kiruna; and finally, the third time period will shed light on the recent developments within foreign- and security policy from 2015 to today.

## 5.1 Foreign Policy Papers from 2003 to 2008 – During Home Rule

From 2003 to 2008 we have six foreign policy papers, and these will give an overview of the priorities that the Government of Greenland had during its time as a Home Rule Government, where limitations in foreign- and security policy were an issue. However, as we have seen in Chapter 4, several attempts towards more influence were made during this period, examples being the Itilleq and Igaliku agreements as well as the Authorization Act. The background for the enhancement of interest within the foreign- and security policy area can be traced back to the work of the Commission of Self-Government in 2000 towards more self-government. The idea was to evaluate the relationship between Greenland and Denmark to modernize the system, hereunder also the foreign- and security policy area (see *Udenrigspolitisk redegørelse 2003*). From the Danish side the Prime Minister, Anders Fogh Rasmussen at the time, was also positive towards more involvement from the Greenlandic side. He meant that it was natural that Greenland should have a say on issues concerning Greenlandic matters within the foreign policy area (*Udenrigspolitisk redegørelse 2003*:39). The priorities in 2003 were "a self-sufficient economy, sustainable development, and partnerships" (*Udenrigspolitisk redegørelse 2003*).

In 2002, a working group had been established to discuss the Defense Agreement from 1951 between the United States and the Kingdom of Denmark. Greenland became involved in the discussions and was also involved in the negotiations regarding a modernization of the agreement, where also Greenland was to become

DOI: 10.4324/9781003387657-5

a co-signatory of the agreement (*Udenrigspolitisk redegørelse 2003*:43). The Greenlandic Representation in Ottawa, Canada was closed in 2002 after being in operation in only a few years from 1998 (*Udenrigspolitisk redegørelse 2003*:47). However, in the recent Arctic Strategy from 2024 to 2033 it is discussed to reopen a Greenlandic representation in Ottawa again (see *Greenland in the World – Nothing about us without us*, 2024:21). During 2002, some activities on the EU-level were implemented. A renewed agreement on the OCT status became implemented and the OCTA was established. Furthermore, Greenland renewed its fishery agreement and established a new Partnership agreement with the EU (*Udenrigspolitisk redegørelse 2003*:54–56).

In 2004 the foreign policy paper focused on the Itilleq agreement and the debate about the environmental clean-up of the Dundas area near Pituffik Space Base. The United States withdraw its responsibility for cleaning up the area, which Denmark then became responsible for (*Udenrigspolitisk redegørelse 2004:6*–9). Another key goal was the ratification of the UN Convention of the Law of the Sea (UNCLOS) from 1982. Denmark ratified the Convention in May 2003 and Greenland also accepted to join the UNCLOS in the autumn of 2003 (*Udenrigspolitisk redegørelse 2004*:9). Continuous cooperation within the Arctic Council and cooperation with Canada and the Nordic countries were also priorities during 2003. The relationship with the EU was highlighted by the debate about the new enlargement of the EU, while ten new members from Eastern Europe accessed the Union in May 2004. For Greenland and the other OCT: s it was of interest how this enlargement would affect the OCT's relationship within the Union (*Udenrigspolitisk redegørelse 2004*:18). In 2004, the foreign affairs changed name to the Directorate of Foreign Affairs with an enhanced staff at the Greenlandic government. From 1 January 2004, Nuuk had a staff with six employees and a secretary, Copenhagen had one employee, and Brussels two employees, one secretary and one intern (*Udenrigspolitisk redegørelse 2004*:37). This shows that foreign- and security policy was an important aspect on the political agenda.

During 2005, the focus was on the Authorization Act, which gave both Greenland and the Faroe Islands possibilities to sign bilateral agreements with foreign countries in areas where they had taken over competencies from Denmark. Another aspect of this was the opportunity to also get involved in international organizations where not only states are members to enhance the Greenlandic and Faroese involvement in foreign affairs (*Udenrigspolitisk redegørelse 2005*:1). Furthermore, the Igaliku agreement was signed on 6 August 2004 at a ceremony in Igaliku in South Greenland. The agreement is an updated version of the Defense Agreement between the USA, Denmark, and Greenland in relation to Pituffik Space Base and it also includes a mutual declaration on cooperation regarding environmental concerns in relation to Pituffik, as well as a mutual declaration regarding technical- and economic cooperation between all three parties (USA, Denmark, and Greenland) (*Udenrigspolitisk redegørelse 2005*:2).

Since Greenland had accepted the UNCLOS in 2004, Greenland's economic exclusive zone became increased to 200 nautical miles. In relation to this a technical investigation also started in relation to the extended continental shelf discussion in relation to the North Pole. It was also agreed with Canada to undertake a technical investigation regarding the Center Line between Greenland and Canada based on modern geodetic data (*Udenrigspolitisk redegørelse 2005*:4).

During the summer of 2004, Greenland also received its first territory underneath the UNESCO World Heritage List and that was the Ilulissat fjord. This had been something which had been underway for several years (since 1996), but the decision came in 2004 to accept the territory to become part of this valuable list (*Udenrigspolitisk redegørelse 2005:*110). Today, a visitor center is located on the heritage site.

Since the Authorization Act in 2005 had become realized it has given Greenland clearer rules of direction of how to act internationally. The Joint Committee between Greenland, Denmark and the USA had its second meeting in Washington in 2005. Within the Nordic cooperation a political debate about the Nordic self-governing territories (the Faroe Islands, Greenland and the Åland Islands) about stronger participation was on the agenda in 2005 during the Danish chairmanship. This had earlier already been a proposal coming from the Faroe Islands back in 2003 (*Udenrigspolitisk redegørelse 2006:*1–3, 9). However, this is still a debate today in 2024 (see for instance KNR-nyheder).

The three autonomous regions would like to become fully-fledged members of the Nordic Council at the same level as the Nordic states. The question has not been solved and there are still discussions going on related to this matter. The Helsinki Agreement of 1962, which has been amended several times, most recently in 1996 needs to be amended. This is now on the agenda of the Nordic Council (see for example: Nordic Co-operation). The focus has become more intensely discussed since Greenland, the Faroe Islands and the Åland Islands were left out from a Ministerial meeting in Finland, where the Arctic and security policy were on the agenda. This has left Greenland with doubt of what the Nordic cooperation serves in relation to Greenlandic affairs (*Udenrigspolitisk redegørelse 2023*:16).

During 2005, Greenland and Denmark disagreed with Canada over the Tartupaluk – Hans Island. Canada had planted a Canadian flag on the Island to show sovereignty over it. After some debates and discussions between the countries, they concluded a declaration of being agreeing to be disagreeing over the matter, but that it should be investigated on the official level for a solution of the problem in relation to technical investigations for upcoming consultations (Udenrigspolitisk redegørelse 2006:8). The disagreement between Denmark and Canada regarding Tartupaluk – Hans Island dates to 1973, since it was never decided how the border should be divided between the countries in relation to the Island. The agreement was only referring to the ocean and the continental shelf between the two countries, but Tartupaluk – Hans Island was left out from the agreement (*Udenrigspolitisk redegørelse 2006:*8).

The year of 2007 has been concentrated around the completion of the Greenlandic Danish Commission's report on the extended self-government for Greenland. The disposition of the foreign policy report for 2007 has changed its design in order to limit the amount of pages and shortened its text without compromising its content (*Udenrigspolitisk redegørelse 2007:*viii, ix).

A focus point during 2007 was the opening of the Air Greenland route from Kangerlussuaq to Baltimore, USA. This was an effort to enhance the possibility of export between the countries and to enhance the cooperation within the educational and cultural area *(Udenrigspolitisk redegørelse 2007:*3). Various projects within different policy fields were also implemented as part of the Joint Committee's umbrella within science and energy, business and tourism, cooperation about national parks, environmental investigations, climate change, language programs in English, telemedicine, etc. *(Udenrigspolitisk redegørelse 2007:*3–4).

In the end of 2005 and beginning of 2006 negotiations about the delimitation of the Center Line in the ocean between Greenland and Svalbard was on the agenda. Greenland participated in the negotiations according to the Itilleq Declaration and the agreement was signed by Norway and Denmark together with Greenland. The agreement was to have the border between both countries' continental shelfs and their fishery zones within the 200 nautical miles (*Udenrigspolitisk redegørelse 2007:*5).

Within the Arctic Council Norway, Sweden, Kingdom of Denmark (incl. Greenland and the Faroe Islands) began to cooperate with the work to have a more long-term vision for the next chairmanships to come after 2006. The work was also to give the Arctic Council a more effective and stable coordination of its work for the years to come. The Government of Greenland was actively involved in the work. Some examples of this being the participation in the Senior Arctic Official meetings and the participation in both the SDWG and CAFF working groups (*Udenrigspolitisk redegørelse 2007:*6). The period of 2007 to 2008 was nominated as the International Polar Year to enhance the international interest and knowledge about the polar regions in relation to the global environment. The Greenlandic led SLiCA project about sustainable living conditions in the Arctic amongst indigenous peoples had also been recognized by the Arctic Council (*Udenrigspolitisk redegørelse 2007:*6–7).

The Nordic cooperation changed character in 2006 after an investigation done by experts of lawyers. The analysis was a result of the earlier mentioned suggestion made by Greenland about the Nordic autonomous regions' participation within the Nordic Council. However, not everyone was satisfied with the result. It became a compromise where the Nordic autonomous regions could participate on an equal basis as their metropolitan countries within the Nordic Council with their own delegations, but they could not take part in the presidium or the control committee. In the Nordic Council of Ministers, it also became possible for the Nordic autonomous regions to take part in committees and working groups, but they could not participate in the voting when decisions were to be made (*Udenrigspolitisk redegørelse 2007:*8–9). This compromise is still under debate today as the autonomous regions are trying to work for more equality within the Nordic cooperation as mentioned earlier.

The relationship between Greenland and the EU was hectic during 2006. New agreements were to be decided upon, which were to replace the Fishery Agreement of 1985. Furthermore, an anti-seal campaign was going on since the EU Parliament had decided that seal products were to be forbidden to enter the EU market due to the Canadian way of harvesting seal meat. The Inuit was to be exempted from this, but it hurt the seal industry and therefore Greenland had to act against this discriminatory decision (*Udenrigspolitisk redegørelse 2007:*11).

In 2008, the preparations of Kingdom of Denmark's chairmanship of the Arctic Council were on the agenda, where also Greenland was to take an active role. The chairmanship period was from 2009 to 2011 (*Udenrigspolitisk redegørelse 2008:*vii).

During the summer of 2007 Russia and Canada had some symbolic acts regarding their requirements over the North Pole. The Greenlandic government stated that Greenland would follow the international rules and regulations in relation to the claims over the extended continental shelf, hereunder UNCLOS (*Udenrigspolitisk redegørelse 2008:*5).

Regarding the Nordic cooperation, a full-fledged membership was not realized after two years of negotiations, however, the Nordic autonomous regions have been given some more room for maneuvering within committees and Nordic institutions. This has an affect especially within the cultural-, educational-, and research area (*Udenrigspolitisk redegørelse 2008:*7). Within the West Nordic cooperation the representatives from the Faroe Islands and Iceland were keen to have Greenland as a member within the Hoyvik-agreement, which is a free trade agreement between Faroe Islands and Iceland. However, the Greenlandic government turned down the invitation but is open for bilateral treaties with respective country (*Udenrigspolitisk redegørelse 2008:*7, see also *Bekendtgørelse af aftale af 31. august 2005 med Island om etablering af et fælles økonomisk område*).

Ilisimatusarfik as a campus was established in 2008 and the university was also becoming a member of *Erasmus University Charter*, which gives students and staff possibilities for exchanges to other European universities and Ilisimatusarfik can also host exchange students and staff from abroad (*Udenrigspolitisk redegørelse 2008:*66).

### *5.1.1 Summary of the Period from 2003 to 2008*

The main objectives of the foreign policy reports have been an enhancement of Greenlandic competences within the foreign- and security policy as we have seen above. Most of the focus has been on legal instruments, such as the Itilleq Declaration, Igaliku Agreement and the Authorization Act. Apart from this an evolvement towards more self-government was also on the agenda during this period. The cooperation of importance is between the EU, USA, Denmark, West Nordic and Nordic countries, but also other countries, such as, Canada. Greenland has also taken an active part in the work of the Arctic Council and within the UN framework.

Common for all six annual reports during the time period from 2003 to 2008 is that all papers are divided into two parts: Part I is concentrated around Greenland's international relations and cooperation with the Kingdom of Denmark, Arctic states and regions, Nordic countries, the EU and the global relations; Part II is handling the issues of resort areas and foreign policy aspects, which have affected political

*Table 5.1* Summary of the Foreign Policy Papers from 2003 to 2008

| *Year* | *Security matters* | *Arctic cooperation* | *ICC* | *Canada* | *Nordic cooperation* | *WNC* | *EU* | *UN* | *WTO* | *Pages* |
|---|---|---|---|---|---|---|---|---|---|---|
| 2003 | X | X | X | X | X | X | X | X | X | 108 |
| 2004 | X | X | | X | X | X | X | X | X | 114 |
| 2005 | X | X | | X | X | X | X | X | X | 120 |
| 2006 | X | X | | X | X | X | X | X | X | 97 |
| 2007 | | X | | X | X | X | X | X | X | 104 |
| 2008 | | X | | X | X | X | X | X | X | 68 |

actions (see *Udenrigspolitiske redegørelser* from 2003 to 2008). In the table below we see the most important areas which have been identified in the documents and the total pages each document consists of (Table 5.1).

The documents contain remarkably similar information, and the descriptions of activities are copy-pasted from year to year with marginal new information. We can see that most areas are included in each paper with exceedingly small diversities between each other. In 2003, the work of ICC is highlighted, while it is lacking in the other documents. The ICC is mentioned also in the other documents, but it does not have a specific section as in 2003. In 2007 and 2008 security matters are not in the forefront, which were seen in earlier documents. However, the work within the Joint Committee is highlighted in these documents.

All documents also describe Greenland's activities in taking part in the various fishery organizations, i.e. NAFMC, NAFO, NEAFC, and NASCO, etc. The bilateral fishery agreements have been updated by the Greenlandic government several times with Russia, Norway, the Faroe Islands, Canada, and Iceland. Greenland has also been active in NAMMCO and IWC (*Udenrigspolitiske redegørelser*, 2003–2008). It is quite clear which policy areas that are the most important ones in a Greenlandic context.

## 5.2 Period from 2009 to 2014 – Introduction of Self-Government

From 2009 to 2014 we have again six foreign policy reports. The government of Greenland has focused on the implementation of the new Self-Government Act that was coming into force on 21st of June 2009. With the enhanced self-government all earlier main principles in relation to foreign- and security policy coming from the Itilleq Declaration and the Authorization Act was included as a separate chapter in the Self-Government Act as Chapter 4 §§11–16 (see Lov nr. 473 af 12/06/2009).

During 2008 some news about CIA flourished, where CIA had presumably undertaken illegal flights over Denmark, Greenland and the Faroe Islands with prisoners onboard. This was highlighted in Danish television during February 2008. The Danish government established a working group to investigate if this had happened. The conclusion was that it was not possible to either validate or reject such an incident. CIA had undertaken civilian flights, but what was onboard the planes was another question. It was not certain that these flights had been used as prisoner transport (*Udenrigspolitisk redegørelse 2009:*13). The result of the event was that the government of Greenland decided to sharpen the control over its airports.

The Ilulissat Declaration was signed 28th of May 2008 by the USA, Russia, Canada, Norway and Denmark (incl. Greenland and the Faroe Islands) regarding the principles to uphold the international regulations by UNCLOS. The countries also confirmed that they would cooperate within research areas regarding the extended continental shelf, the Arctic environment, protection of the vulnerable Arctic environment, preparedness of accidental events and search and rescue (*Udenrigspolitisk redegørelse 2009:*14).

In relation to the EU, 2008 was focused on the implementation of the Partnership Agreement for the period of 2007–2013. Greenland received about 43 million euros each year during that period (*Udenrigspolitisk redegørelse 2009:*18). Another focus was on the "Green book" regarding the future relationships with the OCT countries. The EU was also beginning to focus on Arctic issues in relation to climate change (*Udenrigspolitisk redegørelse 2009:*18–19).

As of January 2009, the Self-Government in Greenland had 13 employees within the foreign- and security area. Ten were working in Nuuk (where one was secretary and another assistant), one in Copenhagen and two in Brussels (*Udenrigspolitisk redegørelse 2009:*30).

Within the transport sector SAS was flying between Kangerlussuaq and Copenhagen during the summer of 2008, and Air Iceland had direct flights between Nuuk and Keflavik. SAS had been given the right to fly between Copenhagen and Kangerlussuaq during the summer period from 2009 to 2011. Air Iceland was eager to extend its flights to also include Narsarsuaq and Ilulissat during 2009 (*Udenrigspolitisk redegørelse 2009:*60).

During 2009, the government of Greenland evaluated Nordic cooperation and developed an overall strategy for the future cooperation. The strategy was based on four main principles:

1 Naalakkersuisut (the Government of Greenland) will take part in the regional cooperation according to the Helsinki agreement.
2 Naalakkersuisut will contribute constructively to the regional cooperation.
3 Naalakkersuisut will use the Nordic cooperation as a platform to enhance Greenlandic interests in other international organizations and institutions.
4 Naalakkersuisut will cooperate actively with the Greenlandic parliamentarians within the Nordic and West Nordic cooperation to be able to influence the common Nordic decisions.

Greenland was an active partner for the Danish chairmanship in 2010 regarding Nordic cooperation (*Udenrigspolitisk redegørelse 2010:*13–14; author's own translation of the main principles). Greenland saw the Nordic cooperation as a steppingstone for further involvement within international relations.

The Greenlandic representation in Brussels began in 2009 to send out a newspaper called "NAP nyt-Bruxelles" with information about events that had occurred in Brussels and news about new negotiations and agreements with the

EU (*Udenrigspolitisk redegørelse 2010*:20). This was based on the idea from the Danish Foreign Ministry, which had a similar newsletter. The newsletter was spread to interested parties and therefore more widespread knowledge about the EU matters was outlined also for the public.

During the autumn assembly in the Greenlandic parliament in 2009, Inatsisartut (the Greenlandic parliament) adopted the Haag Convention from 1993 regarding protection of children in relation to international adoptions (*Udenrigspolitisk redegørelse 2010*:33). This had been a matter that had been debated in the Greenlandic parliament and now the implementation of the international convention became realized.

The Kingdom of Denmark including Greenland, and the Faroe Islands took over the chairmanship of the Arctic Council in April 2009 for the period of 2009–2011. Greenland has been active in several working groups and been able to promote Greenlandic interests within the Arctic cooperation (*Udenrigspolitisk redegørelse 2011*:12). Greenland had the role of chair for some of the working groups and was therefore involved in several projects underneath the Arctic Council. Within the Arctic Council, the first binding agreement regarding search- and rescue missions[1] (so-called SAR operations) became realized at the Ministerial meeting in Nuuk, Greenland in May 2011. Another issue that was on the agenda was to establish a permanent secretariat in Tromsø for 2013 (*Udenrigspolitisk redegørelse 2012*:13).

The Kingdom of Denmark's Arctic strategy from 2011 to 2020 became launched with four main themes: "a peaceful Arctic, selfsustained economic growth and societal sustainability, development regarding respect of the Arctic climate, environment and nature, as well as a close collaboration with international partners" (*Udenrigspolitisk redegørelse 2012*:10). These overall themes were also followed by the Government of Greenland. The Kingdom of Denmark's Arctic strategy from 2011 to 2020 was focused on the relationship between Nuuk and Copenhagen in particular.

During 2012, Russia was implementing prohibition towards import of sealskin from Greenland. The Foreign Ministry from Denmark, Department of Hunting, Fishery and Agriculture from Greenland, and the EU was trying to persuade the Russians to accept the exemption of sealskin import from indigenous peoples as the EU had done. Greenland was not notified about this prohibition. This had severe consequences for Great Greenland in Russia, since this was one of the largest markets for sealskin after the EU market (*Udenrigspolitisk redegørelse 2012*:30).

Greenland signed the first Nordic Convention on Social Insurance on the 12th of June 2012 as an independent nation. The Convention was based on EU-regulation nr. 883/2004 about coordinating social insurance systems. This Convention includes issues related to pensions, family services, some health services and unemployment support (*Udenrigspolitisk redegørelse 2012*:56). The ratification was planned to be done in the autumn of 2012.

Greenland signed a cooperation agreement with the European Environment Agency in January 2012. This was to give Greenland possibilities to exchange knowledge about how the environment is affecting the health of the population. This also led to that investigations would be available in Greenlandic and English (*Udenrigspolitisk redegørelse 2012*:58).

In August 2012, there were some discussions regarding extraction of uranium and rare earth elements and its consequences in relation to foreign- and security policy. This led to an agreement between the Government of Denmark and the Government of Greenland regarding the extraction of uranium and other radioactive substances, which was implemented in 2015 (*Udenrigspolitisk redegørelse 2013*:5; *Aftale mellem den danske regering og Naalakkersuisut vedrørende de særlige udenrigs -, forsvars- og sikkerhedspolitiske forhold, der knytter sig til udvinding og eksport af uran og andre radioaktive stoffer i* Grønland, 2015).

It became clear during 2013 that Greenland was going to open a new representation in Washington D.C. This had been underway for a few years, but the representation opened in 2014. The representation is in the same building as the Danish embassy. Now the Greenlandic representation is also accredited to Canada (*Udenrigspolitisk redegørelse 2013*:5).

Sweden took over the chairmanship in the Arctic Council in 2011–2013. One of the major tasks was to establish the permanent secretariat in Tromsø and draft the agreement regarding oil spills in the Arctic, the *Agreement on Cooperation on Marine Oil Pollution Preparedness and Response in the Arctic*, which was the second binding agreement within the auspices of the Arctic Council, even though it is not part of the policy to have binding agreements in this forum. Furthermore, six new observer applications were accepted at the Kiruna meeting in 2013 (*Udenrigspolitisk redegørelse 2013*:8).

The Swedish chairmanship sparked some issues regarding the leadership management and Greenland was therefore boycotting the Kiruna meeting in 2013. The Swedish approach was to only accept member states around the negotiation table, excluding the Faroe Islands and Greenland, which had been sitting around the table together with Denmark. The "missing chair-policy" became criticized from Greenland and therefore Aleqa Hammond, who then was the Greenlandic premier at the time, suspended all cooperation with the Arctic Council in hope of that a solution was to be reinforced. (*Udenrigspolitisk redegørelse 2013*:8; see also earlier section about the issue). With the Canadian chairmanship 2013–2015 a solution was reinforced as a compromise, which was acceptable for Greenland. Now, there is a rotating procedure between the representatives from the whole Kingdom of Denmark; each party has one minute each to make their statements. Another issue was also to move the Indigenous Peoples' Secretariat to Tromsø, since that would be more practical in collaboration with the now implemented permanent Arctic Council Secretariat. This was accepted by all parties (*Udenrigspolitisk redegørelse 2013*:9).

The cooperation between Greenland and Canada was leading to a new agreement regarding the delimitation of the continental shelf between the countries. This was agreed upon in November 2012 and was an adjustment of the agreement from 1973 based on modern technology and an extension of the northern border. This agreement is not including Tartupaluk – Hans Island (*Udenrigspolitisk redegørelse 2013*:11).

The Nordic cooperation has its own Nordic strategy, but there has been a development of a cooperative forum for Faroe Islands, Greenland and the Åland Islands, which means that the idea is to work together and strengthen each other in finding solutions within the Nordic cooperation. The declaration has been signed by the three Nordic autonomous regions and the delegation leaders from all Nordic states. This happened in May 2013 (*Udenrigspolitisk redegørelse 2013:*12).

Within the EU a new fishery agreement was struck between Greenland and the EU and other agreements were also updated, such as the common political declaration, and the OCT status was renewed for the period of 2014–2020. The estimated economic support from the EU is about 320 million DKK (ca. 43 million €) annually throughout the period (*Udenrigspolitisk redegørelse 2013:*14).

A new bilateral agreement was struck with South Korea during 2012. Cooperation between Greenland and South Korea is focused on education and business development. A free trade agreement is also underway and is expected to be implemented in 2013 (*Udenrigspolitisk redegørelse 2013:*30).

Iceland opened its consulate in Nuuk in November 2013 to enhance its cooperation with Greenland and as already mentioned the Greenlandic representation in Washington D.C. opened in January 2014. Other top events during 2014 were the negotiations about an agreement of high seas fishing in Nuuk and a visit by the UN General Secretary (*Udenrigspolitisk redegørelse 2014:*3).

There has been a discussion to open a new representation in Beijing, China for 2015 (Udenrigspolitisk redegørelse 2014:3). However, this representation has been delayed and was realized during the pandemic in 2021 (see later sections).

The Joint Committee between Greenland and the US has been discussing issues concerning language and education, research and economic cooperation during 2013. The results of the negotiations have been that the English courses at the business school have been doubled and in Qaqortoq tourism and English have been developed at the business school there. Furthermore, the Fulbright director for the commission of Denmark-USA foundation has visited Greenland (*Udenrigspolitisk redegørelse 2014:*5). This has opened for more knowledge in the English language and possibilities for exchange in North American universities for higher education.

During the Canadian chairmanship of the Arctic Council 2013–2015, the Arctic Economic Council (AEC) was established. The idea behind this forum was to enhance the cooperation within the business sector in the Arctic and to support the Arctic Council with knowledge about economic matters related to the development in the Arctic overall. Greenland was now also participating in the Arctic Council again (*Udenrigspolitisk redegørelse 2014:*7).

Greenland was actively involved in the preparation for the Danish chairmanship of the Nordic Council for 2015. The program focused on growth, welfare and values. A specific section was devoted to Arctic issues in relation to the ocean and its resources (*Udenrigspolitisk redegørelse 2014:*11).

### *5.2.1 Summary of the Period 2009–2014*

The highlights from the period of 2009–2014 are the implementation of the Self-Government Act and by that Greenland has taken more room of maneuver in the international sphere by signing conventions and agreements with foreign countries and international organizations. The Arctic Council has been developing

*Table 5.2* Summary of the Foreign Policy Papers 2009–2014

| ***Year*** | ***Security matters*** | ***Arctic cooperation*** | ***Canada*** | ***Nordic cooperation*** | ***EU*** | ***UN*** | ***WTO*** | ***Pages*** |
|---|---|---|---|---|---|---|---|---|
| 2009 | X | X | X | X | X | X | X | 86 |
| 2010 | | X | X | X | X | X | X | 77 |
| 2011[a] | | X | X | X | X | X | X | 25 |
| 2012 | | X | X | X | X | X | X | 67 |
| 2013[b] | X | X | X | X | X | X | | 62 |
| 2014 | X | X | X | X | X | X | | 55 |

[a] The 2011 foreign policy paper is in another format than the rest of the foreign policy papers. It is in a shorter format with some visions for the future.

[b] From 2013 and onwards the foreign policy papers have pictures, but otherwise it is the same format as earlier with two parts.

to become a stronger organization with its own permanent secretariat and we also see an enhancement of the cooperation within certain areas leading to binding agreements, furthermore Greenland has taken active part in developing its own representation abroad with opening a new office in Washington D.C. and we see that Iceland has opened its consulate during the period to enhance cooperation with Greenland. Added to this we witness more visits from abroad from other countries and the UN. All in all, this period has been rather active and productive (Table 5.2).

The foreign annual reports from 2009 to 2014 have seen a little bit of a shift. They are all, apart from the annual report from 2011, according to the same structure as before, i.e. divided into two parts. From 2013 pictures are included and a new layout is used to make the reports more readable. West Nordic cooperation is not mentioned in the reports from this period, it is only Nordic cooperation overall that is mentioned. This might be because there are separate Nordic strategies and reports also produced by the Government of Greenland, so information has been limited to the main activities in the reports under investigation. Again, all the fishery organizations and mammal organizations are mentioned in all foreign policy reports.

## 5.3 Period from 2015 to 2023 – Enhancement in IR

The period from 2015 to 2023 gives us nine foreign policy reports. The policy reports with the exemption of the annual report from 2023 are all divided into the previous two parts.

One of the intriguing issues during 2014 was the loss of the service contracts at the former Thule Air Base (now Pituffik Space Base). On the 31st of October 2014 the USA gave a statement that the contract was to be given to the Exelis company (later renamed to Vectrus Services). This company was seen as a shell company from the Danish/Greenlandic side and was seen as a breach on the Defense Agreement between the three parties (USA, Denmark and Greenland). The issue became a heated debate on high diplomatic and political level and the governments in Denmark and Greenland were trying to secure the contracts to a Danish/Greenlandic company. The Danish/Greenlandic company of Greenland Contractors, which had the contract before raised a complaint and a court case in the USA. The decision came in 2015, and the verdict was that Exelis would be given the contracts, however Greenland Contractors were given support in the case (*Udenrigspolitisk redegørelse 2015:*15). In the end, Exelis came out as the winner of the dispute, but the issue was to be solved for the next round of service contract negotiations.

During the Canadian chairmanship of the Arctic Council and the Ministerial meeting in Iqaluit, Nunavut, 24th of April 2015 the Foreign Minister from Greenland participated for the first time in the informal meetings after the earlier mentioned boycott. During these meetings several issues were discussed, such as, the status of the EU as an observer, populations in the Arctic, the balance of environmental protection and sustainable economic development and other focus areas of vital importance for the years to come (*Udenrigspolitisk redegørelse 2015:*17). The USA took over the chairmanship after Canada in 2015.

Greenland participated in May 2015 in the UN discrimination against race examination in Geneva together with Denmark. The themes that were relevant for Greenland were the following:

- The issue of general legislation regarding discrimination against race
- Establishment of data regarding race matters and added to this an institution of complaint regarding these issues
- The issue of the juridical fatherless individuals
- The case of the Thule population
- The Greenlandic language and the Greenlanders living in Denmark (Udenrigspolitisk redegørelse 2015:29).

All UN member countries, hereunder Greenland as a part of the Kingdom of Denmark is going through a Universal Periodic Review every 4th or 5th year. This is to check if the countries and nations are living up to the Human Rights obligations and the conventions they have signed in relation to the UN (*Udenrigspolitisk redegørelse 2020:*52).

Within the sector of natural resources there have been some bilateral agreements that have been struck between Greenland and Canada and Norway in 2015. With Norway one agreement is with the Norwegian Oil Directorate and with Canada it is with the National Energy Board. There has also been an agreement with the Norwegian Petroleum Agency. The Department of Mineral Resources in Nuuk has also had some exchange of experiences with American authorities, such as, Bureau of Safety and Environmental Enforcement and Bureau of Ocean Energy Management. Added to this the Department of Mineral Resources in Nuuk has signed a Memorandum of Understanding with South Korea regarding natural resources (*Udenrigspolitisk redegørelse 2015:*51). This shows that the Greenlandic authorities try to receive information and exchanges of experiences from other countries and in that vein try to build up best practices for the industry.

Brexit became a large event in 2016, where the Brits voted to leave the EU. This has had some consequences also for Greenland, since United Kingdom is one of the largest markets for shrimps and cod (*Udenrigspolitisk redegørelse 2016:*9). Another event, which had some interest during the end of 2016 was the presidential election in the USA, which saw a new era of a new type of administration underneath Trump (*Udenrigspolitisk redegørelse 2016:*10).

During 2016, Greenland was also signing a MoU with China between the Department of Education, Culture, Research and Church and China's State Oceanic Administration and the Polar Research Institute in China. The purpose with this MoU was to establish closer cooperation between Greenland and China in relation to Arctic research, for example in relation to research stations and for exchange of researchers and students between the two countries. Department of Foreign Affairs have an advisory role in this cooperation as well as within the work underneath the Scientific Cooperation Task Force under the Arctic Council (*Udenrigspolitisk redegørelse 2016:*15).

Public diplomacy became a new activity, which was launched in 2016. This idea was to enhance the information of Greenland within international debates. The purpose was to use social media to interact with governments, politicians and officials, but also with civil society. The idea was to have a platform where Greenlandic voices could be heard and where Greenland could have some influence over other decision-makers. Another purpose was to enhance Greenland in a positive light in order to receive more foreign direct investments, have better opportunities for export and attract new tourists (*Udenrigspolitisk redegørelse 2016:*18).

Greenland and Denmark had some discussions about cooperation regarding the uranium, issues in relation to the extended shelf, Greenlandic products entering export markets, issues related to the International Whaling Commission (IWC), and EU and Arctic Council matters. Added to this cooperation regarding the American base was discussed (Udenrigspolitisk redegørelse 2016:20). The Danish Government had decided in 2013 to start a large analytical work regarding the Ministry of Defense in Denmark and its work in the Arctic. A report was presented on 27 June 2016 in Copenhagen, where also the Greenlandic Minister of Foreign Affairs at that time participated. The result of the report was that the Arctic should be prioritized, and Greenland had also been involved in the work of the report. Some proposals were to establish a voluntary unit in Greenland and to enhance the education within search and rescue (*Udenrigspolitisk redegørelse 2016:*25).

The government of Greenland and the government of Denmark signed an agreement regarding the uranium issue in January 2016. This was based on an earlier report, which had been launched in 2013 regarding the issue of entering an agreement complex between Denmark and Greenland of future cooperation in relation to exploitation and export of uranium from Greenland. Denmark was to be responsible for non-proliferation of nuclear weapons and material, hereunder also safety measures and security and export of dual use products. The agreement emphasizes the common Danish and Greenlandic responsibility of upholding the highest international standards in relation to other uranium exporting countries. The Danish legislation for Greenland is to secure that the export is correctly done and according to international standards and that this is secured for peaceful purposes (*Udenrigspolitisk redegørelse 2017:*74–75).

The Danish Government and the Greenlandic Government were sending their part submissions around the extended continental shelf in 2016. There are three areas in Greenland, which are of interest regarding the extended continental shelf. The submissions are sent to the Commission on the Limits of the Continental Shelf – CLCS to the UN headquarters in New York. A delegation from Greenland and Denmark visited New York in August 2016 to state their requirements. It will take years before a decision is made regarding this issue. It will then be up to the states to agree on overlapping requirements and negotiate delimitation agreements (*Udenrigspolitisk redegørelse 2018:*28).

The Arctic Council was celebrating its 20th anniversary on 16th September 2016. The USA had taken over the chairmanship of the Arctic Council from Canada in April 2015. The US chairmanship was focusing on maritime environment, sustainable energy, infrastructure and the welfare of the populations in the Arctic (*Udenrigspolitisk redegørelse 2016:*27).

Greenland, Denmark and the EU was signing a Joint Declaration to have an overall overview of the various agreements and their obligations for the partnership between Greenland and the EU on 19th of March 2015. The Joint Declaration was implemented in 2016 (*Udenrigspolitisk redegørelse 2016:*35). The EU was also implementing its own Arctic strategy at the time.

The Arctic Council signed a new binding agreement in 2017 around cooperation of science, the so-called Agreement on Enhancing International Arctic Scientific Cooperation. Greenland was also signing this agreement (Udenrigspolitisk redegørelse 2017:7). Furthermore, new observers were accepted to join the Arctic Council at the ministerial meeting 11th of May 2017; ICES (International Council of the Exploration of the Sea), the National Geographic Society, Oceana, OSPAR, Switzerland, the West Nordic Council and the World Meteorological Organization all joined as observers (*Udenrigspolitisk redegørelse 2017:*21). Greenland was together with Canada co-leading the SDWG-project "Rising-Sun" about suicide prevention during the American chairmanship of the Arctic Council (*Udenrigspolitisk redegørelse 2017:*54). Greenland was furthermore an active participant in the AMAP working group, which had a focus on environmental issues and climate change in relation to environmental toxins in danger for health amongst both humans and animals. In April 2017, a new report "Snow, Water, Ice, Permafrost in the Arctic" was launched about the climate changes in snow, ice and water conditions in the Arctic. Another report was also launched in 2017 called "Adaptation Actions for a Changing Arctic". This report was focused on the challenges and possibilities that climate change gives in several sectors in West-Greenland and Nunavut (*Udenrigspolitisk redegørelse 2017:*80). Greenland has been Head of Delegation also for CAFF working group and been participating in PAME's meetings (*Udenrigspolitisk redegørelse 2017:*81).

Another event during 2017 was the realization of Brexit. The referendum had been on 23rd of June 2016, but the realization of the process of leaving the EU was postponed until March 2017. It will take two years to negotiate the Brexit (*Udenrigspolitisk redegørelse 2017:*31).

The situation of Russia was discussed during an EU-Arctic High-Level Event in Finland on 15th of June 2017. Russia's participation in the Arctic cooperation was questioned due to the security situation in the Arctic, however the speakers were agreeing on that the Arctic cooperation is a peaceful cooperation. Focus should be on education, infrastructure, and securing equality, stability and development in the region (*Udenrigspolitisk redegørelse 2017:*34).

During 2017, the Nordic Council had the issue of full-fledged membership for Greenland up on the agenda, however no decision was made, and this is still an ongoing debate (*Udenrigspolitisk redegørelse 2018:*29).

A new Greenlandic representation opened its doors in Reykjavik, Iceland on 20th of October 2018. This was to enhance and strengthened the cooperation with the nearest neighbor country (Udenrigspolitisk redegørelse 2018:11). In 2019, the representation in Reykjavik was developed to include a secretary and an internship program in the same way as has been done in Brussels and Washington D.C. (*Udenrigspolitisk redegørelse 2019:*14).

The Ilulissat Declaration from 2008 had its ten-year jubilee and was celebrated with a ceremony in Ilulissat in Greenland during 22–23 May 2018. All Arctic states and permanent members of the Arctic Council participated in this ceremony. Back in 2008 it was only the coastal Arctic-five states that participated in the Ilulissat Declaration, but now all Arctic states were invited to become included in the process of implementing the principles laid down in the document (*Udenrigspolitisk redegørelse 2018:*13).

In the beginning of 2018, the European Commission presented a report regarding investments in the Arctic based on input from the Arctic states. This report will be integrated into the future budget negotiations taken place within the EU for the upcoming period of 2021–2027 (*Udenrigspolitisk redegørelse 2018:*41).

The cooperation with the USA and Canada has become more enhanced and seen as important for Greenland. In February 2018, Greenland participated in the Northern Lights conference in Ottawa, Canada. This conference has the purpose of portraying the Arctic regions of Nunavut, Nunavik and Labrador/Nunatsiavut and their businesses, industry, history and culture. Greenland was invited to present the possibilities of cooperation with the Canadian regions (*Udenrigspolitisk redegørelse 2018:*50). Greenland has also listed a long list of cooperation possibilities with the USA for the years to come. Amongst these are to enhance the knowledge of Greenland, establishing network within businesses, agreements regarding education and research, exchange opportunities between Greenlandic and American institutions, establishing internship programs with companies and educational programs at universities and so on (*Udenrigspolitisk redegørelse 2018:*48).

In 2018, it was also the first time that the Danish defense offered a test for conscription for the military for Greenlanders in Greenland for those who volunteer in the army. The basis education is however taken place in Denmark. In 2019 it is estimated that the test of conscription will take place in four places in Greenland: Nuuk, Sisimiut, Aasiaat and Qaqortoq. All towns have gymnasiums (*Udenrigspolitisk redegørelse 2019:*19).

In 2019, the government of Greenland and the government of Denmark signed an agreement regarding the defense use of the airport in Kangerlussuaq after 2023, when the new airport in Nuuk is going to open (has opened on 28th of November 2024). The Danish defense has therefore the possibility to use the airport also in the future and it is also possible for civilian aircraft to use it. The Greenlandic ownership will not change and there is an understanding that the Ministry of Defense in Denmark will manage the airport, i.e. undertake the costs.

In the Arctic Council it was the first time during a ministerial meeting in May of 2019 underneath the Finnish chairmanship that the Arctic states did not came to an agreement about the ministerial declaration due to the new Trump administration in the USA. The wording of climate change and global warming was not agreed upon. However, IMO became a new observer. Iceland was taken over the chairmanship after Finland with a focus on the Arctic Ocean, climate and green transition, people and society in the Arctic, and a stronger Arctic Council (*Udenrigspolitisk redegørelse 2019:*20).

The rights of indigenous peoples were one of the key priorities during 2019. The government of Greenland nominated together with the government of Denmark a Greenlandic member to the UN Permanent Forum on Indigenous Issues (UNPFII) for the period of 2020–2022 (*Udenrigspolitisk redegørelse 2019:*44). The 13th of April 2022, the Greenlandic representative became re-elected for the period of 2023–2025 *(Udenrigspolitisk redegørelse 2022:*44).

The relationship with the EU was also an important topic on the agenda, since the political joint declaration, the OCT- agreement, the partnership agreement, and the fishery agreement were going to end in 2020. That meant that new negotiations needed to start between the EU Commission and Greenland and Denmark (*Udenrigspolitisk redegørelse 2019:*12–13).

Brexit became reality on the 31st of October 2019. There had been a lot of disagreements of how to leave the EU from the British side. In the end it became a "hard Brexit" meaning that the United Kingdom was leaving without an agreement with the EU. Some of the areas will have a transition period until the end of 2020. An old agreement from 1959 between the United Kingdom and Denmark (including Greenland and the Faroe Islands) came into force again regarding social security, which means that citizens from Greenland can receive the same health care services as British citizens if necessary and vice versa (*Udenrigspolitisk redegørelse 2019:*29–30).

A highlight for Greenlandic shipping became realized in 2019, when Royal Arctic Line A/S and Eimskib opened a shipping route to Portland in Maine in the USA. Maine has already an established cooperation with Iceland through Eimskib so therefore it was a natural extended cooperation with also Royal Arctic Line A/S (*Udenrigspolitisk redegørelse 2019:*40). However, there are still a lot of work to be done to reduce the trade barriers between Greenland and the USA and Canada. One example of this can be illustrated by Royal Arctic Line A/S sending five boxes of water through Reykjavik in Iceland to a West Nordic reception to be held on 22 September 2023 in Washington D.C., where each bottle of water had a price tag of DKK 111 per bottle (about US$ 16/bottle), which is of course not sustainable and a challenge for trade between the two countries (*Udenrigspolitisk redegørelse 2023:*29).

The Government of Greenland was also beginning to see the opportunity to open a representation in Beijing, China for 2020. This would strengthen the trade, business and political cooperation between Asian countries and Greenland not just China (*Udenrigspolitisk redegørelse 2019:*43). However, because of the

COVID-19 restrictions in China, the official opening of the representation was realized in 2023 (*Udenrigspolitisk redegørelse 2023:*32).

A global pandemic, the COVID-19 crisis, was the topical issue during 2020–2021, which meant that several high diplomatic meetings and conferences were moved online, cancelled or moved to a later date. This affected a lot of cooperation within the Arctic Council, UN, EU and bilateral relations (*Udenrigspolitisk redegørelse 2020:*23; Udenrigspolitisk redegørelse 2021). Greenland was however not hit that hard, since it was possible to close the borders and the pandemic came with a delay (see more information Ackrén, 2024:105–112).

Greenland did work on a new negotiation round with the EU during 2020. The year of 2020 was the last year valid for the Partnership Agreement, so a new agreement was therefore worked on in order to secure the period of 2021–2027 (*Udenrigspolitisk redegørelse 2020:*34). Greenland has a good track record when it comes to favorable agreements with the EU, which is an amazing achievement of such a small nation.

The State Department of the USA decided to reopen the Consulate in Nuuk in June 2020. The State Department has also decided to invest DKK 82 million (US$ 12.1 million) in Greenland for development within tourism, the mining sector and education. It was also a wish from the politicians to reactivate the meetings within the Permanent Committee, which had not had any meetings since 2014 due to the dispute of the service contracts in relation to Pituffik Space Base (*Udenrigspolitisk redegørelse 2020:*43; Udenrigspolitisk redegørelse 2021:40). The complex agreement package in relation to Pituffik Space Base came to an end on 28th of October 2020. Greenland will in the future have more say and the slogan "nothing about us without us" became articulated. There are four documents related to Pituffik Space Base. One diplomatic note about the new rounds of contract tender criteria; an agreement about initiatives for cooperation in Pituffik; a statement about improved cooperation; and a declaration about enhancing the existent areas of cooperation between the USA and Greenland. The declaration includes cooperation in trade, investments and economic development, cooperation within energy and the mineral industry, cooperation within education, tourism and natural environment management (*Udenrigspolitisk redegørelse 2021:*40). The tender criteria include regulations that the company should be registered as a Danish or Greenlandic company, more than 50% of own capital should be owned by Danish and/or Greenlandic private persons or jurisdictional entities, a non-Danish or non-Greenlandic person or jurisdictional entity cannot have a crucial influence on the providers (*Udenrigspolitisk redegørelse 2021:*41).

The think tank, Woodrow Wilson Center in Washington D.C., has established a Polar Initiative and the idea is to enhance the cooperation between the USA and Greenland through various activities such as seminars, workshops and research. In this way the American public will become more knowledgeable about Greenland and its self-government and exchanges of knowledge between the two countries will enhance the diplomatic relationships and relationships between researchers (*Udenrigspolitisk redegørelse 2021:*43). Another initiative has been established at Harvard Kennedy School in Boston, Massachusetts where a program called Arctic Initiative has been launched. The Arctic Initiative is a program where new insights and cooperation ties science, technology and political strategies together. The idea is to come up with solutions for the Arctic region in combatting the social, economic and environmental challenges that the region faces (*Udenrigspolitisk redegørelse 2021:*43).

A new Arctic strategy was supposed to be discussed within the Kingdom of Denmark during 2020, including the Faroe Islands and Greenland, however this has been postponed. Greenland was stating that they would launch their own strategy first. This happened in the Spring of 2024 and there is still no joint Arctic strategy for the whole Kingdom.

Within the Nordic cooperation the Kingdom of Denmark had the chairmanship during 2020. The Nordic vision of 2030 was given a major role in the program. Faroe Islands was given the chairmanship of the fishery sector, while Greenland was taken care of the development of sustainable coastal regions in the Nordic countries (*Udenrigspolitisk redegørelse 2020:*71).

The Government of Greenland is participating in the Arctic Offshore Regulators Forum (AORF). The AORF is a cooperative forum with members of the offshore authorities in the USA, Canada, Russia, Iceland, Greenland, Finland, Sweden and Norway. The purpose of the forum is to cooperate around security issues related to the challenges and possibilities of offshore activities in the Arctic. Within the forum exchanges of experiences and knowledge is done across the authorities within this field (*Udenrigspolitisk redegørelse 2020:*81). From 2021, the new Government of Greenland is stating that it will forbid future exploitation of uranium and put an end on future oil exploration and exploitation around Greenland (*Udenrigspolitisk redegørelse 2021:*12).

In cooperation with Danish and Greenlandic authorities within research an International Arctic Hub was to be established in 2020. The Arctic Hub is a platform in Greenland for Danish, Greenlandic and international polar researchers and Arctic interests (*Udenrigspolitisk redegørelse 2020:*82). Today, the Arctic Hub is very active in organizing seminars, workshops and the PhD-cup (*Paasisavut*). The purpose is to be a platform for connecting people, institutions and countries, as well as building bridges between research and society (for more information visit the webpage: Arctic Hub).

With enhanced interest in the Arctic, it has been evaluated that the risk of cyberattacks in Greenland will increase. The Danish Ministry of Defense and the Danish Intelligence Service (PET) was launching a report in 2019 where the Arctic became on top of the list for the first time. The risk analysis describes the great power rivalry in the Arctic. Authorities and businesses on a world stage are increasingly becoming subjects of cyberattacks (*Udenrigspolitisk redegørelse 2020:*93). There have been some incidents also in Greenland, where cyberattacks have been targeted towards the Greenlandic Government, shipping and the health sector.

In the end of 2020, the West Nordic Council was declaring that the Council was supporting Faroe Islands and Greenland to become associated members of the World Health Organization (WHO) (Udenrigspolitisk redegørelse 2021:29). Denmark on behalf of the Faroe Islands sent an application for associated membership on 18th of May 2021 to the WHO (WHO: Application: https://apps.who.int/gb/ebwha/pdf_files/WHA74/A74_INF5-en.pdf).

A new challenge which made headlines throughout the world was Russia's invasion of Ukraine on 24th of February 2022. Greenland was connecting with the international society's condemnation of Russia's illegal acts and breach of international law. Greenland has furthermore signed the international sanctions towards Russia. Due to Russian war against Ukraine all other Arctic states paused their cooperation within the Arctic Council. The Arctic Seven (excluding Russia) has tried to figure out how the cooperation can continue in the Arctic Council, since it is important for the countries to carry out their work (Udenrigspolitisk redegørelse 2022:10). Since June 2022, the Arctic Seven reestablished a limited way to work within the Arctic Council so the six working groups at least could carry out their tasks. In May 2023, the chairmanship changed from Russia to Norway through an online ceremony. The Arctic Council has reestablished its work through written procedures and online meetings. In May 2025, the Kingdom of Denmark will take over the chairmanship and Greenland will play a key role in the work, since Greenland is the Arctic part of the Kingdom (*Udenrigspolitisk redegørelse 2023:*6–7).

Due to the changing character of world politics in relation to defense and security, Greenland decided to send a Greenlandic diplomat to the Danish permanent representation to the NATO headquarters in Brussels in 2023. The Greenlandic diplomat is included as part of the Danish diplomatic corps and can contribute with their expert knowledge of Arctic conditions and therefore advance the Kingdom of Denmark's security priorities (*Udenrigspolitisk redegørelse 2023:*11).

A highlight in the cooperation with Canada was the signing of the agreement regarding Tartupaluk – Hans Island between Denmark, Greenland and Canada on 14th of June 2022. This agreement is an agreement with the longest maritime border in the world of 3,926 km and one of the shortest land borders in the world with only 1.2 km. It is the first time that the Government of Greenland has been working with a border treaty. In relation to this agreement, it is suggested that Greenland and Canada take the Inuit free movement into account and that joint regulations for this should be implemented between the two countries (*Udenrigspolitisk redegørelse 2023:*13). This is also an example on that it is possible to reach an agreement in a peaceful way when it comes to border issues. The dispute has been a long-lasting diplomatic dispute since 1973 (for more information see Lackenbauer and Nielsen, 2022).

The Greenlandic representation in Brussels changed officially name to Greenland's Mission to the EU in Brussels during 2022 (Udenrigspolitisk redegørelse 2022:15). Cooperation with the EU is important and the yearly support from the EU is about DKK 379 million (€51 million). A large proportion of the financial support is the expenses for fishing rights in Greenlandic waters. The new budget is running from 2021 to 2027. There are also focus on education and green growth were DKK 1.6 billion (€225 million) is going to Greenland for the mentioned period (Udenrigspolitisk redegørelse 2022:27). An EU Commission office was also established in Nuuk and opened its doors in the Spring of 2024. This has been a negotiation going on since May 2022 and in January of 2024 it was agreed that the new office was to become a reality (*Udenrigspolitisk redegørelse 2023:*34).

A Canadian company has won the competition to supply the upcoming new airport in Qaqortoq in South Greenland 2022. It is Pennecon from Newfoundland, which will build the airport and has started the project in the Spring of 2022. It is estimated that the new airport will open in 2025 (*Udenrigspolitisk redegørelse 2022:*38). Air Greenland and Canadian North has signed a MoU around cooperation within aviation and tourism development. Future air connections will potentially be between Nuuk/Ilulissat and Toronto/Quebec (*Udenrigspolitisk redegørelse 2022:*39). During the Summer of 2024 Air Greenland and Canadian North were cooperating with a reopened route between Nuuk and Iqaluit in Nunavut (*Udenrigspolitisk redegørelse 2023:*27). It is estimated that this route will continue during the summer of 2025.

The airspace of the higher airspace over 19,500 feet is controlled by Iceland and Canada. The Government of Greenland would like to take over the airspace control. It is proposed that the Government will work towards this, and a control center will be established, so Greenland can begin to operate in 2031 and have control over its own airspace (*Udenrigspolitisk redegørelse 2022:*54). This is in line with the new international airport, which has recently opened in Nuuk and the extension of the airport in Ilulissat and the new airport as mentioned above in Qaqortoq.

In December of 2022, *Qilaatersorneq* (drum dance and song) became part of UNESCO's representative list over immaterial culture heritage. This was decided at the Intergovernmental Committee's 16th session in Paris. *Qilaatersorneq* is the first Greenlandic immaterial cultural heritage element, which has been listed so far (*Udenrigspolitisk redegørelse 2022:*85).

### *5.3.1 Summary of the Period 2015–2023*

The last nine years have been turbulent years with a worldwide pandemic, Russian invasion of Ukraine, BREXIT, disputes about service contracts in relation to Pituffik Space Base, two new representations abroad (Reykjavik and Beijing), a peaceful solution on the Tartupaluk – Hans Island dispute, a reopening of a US Consulate in Nuuk, new negotiations with the EU, etc. There have been both challenges and opportunities in a Greenlandic context. With an extended Department of Foreign Affairs, we see more interactions with foreign countries and more cooperation in various fields. New agreements are struck, and new aims are set for future purposes. All foreign policy reports are concentrated around the same issues as can be seen from the table below (Table 5.3).

*Table 5.3* Summary of the Foreign Policy Papers 2015–2023

| *Year* | *Security matters* | *Arctic cooperation* | *Nordic cooperation* | *EU* | *UN* | *USA* | *Canada* | *Asia* | *Pages* |
|---|---|---|---|---|---|---|---|---|---|
| 2015 | X | X | X | X | X | X | X | X | 75 |
| 2016 | X | X | X | X | X | X | X | X | 108 |
| 2017 | X | X | X | X | X | X | X | | 84 |
| 2018 | X | X | X | X | X | X | X | X | 93 |
| 2019 | X | X | X | X | X | X | X | X | 85 |
| 2020 | X | X | X | X | X | X | X | X | 95 |
| 2021 | X | X | X | X | X | X | X | X | 94 |
| 2022 | X | X | X | X | X | X | X | X | 88 |
| 2023[a] | X | X | X | X | X | X | X | X | 40 |

[a] 2023 Foreign Policy Report is in a totally new format with only the Ministry of Statehood and Foreign Affairs. It lacks resort areas from the other departments.

## Note

1 The agreement is called "Agreement on Cooperation on Aeronautical and Maritime Search and Rescue in the Arctic", see https://oaarchive.arctic-council.org/items/9c343a3f-cc4b-4e75-bfd3–4b318137f8a2 (Visited 4 October2024).

## References

Ackrén, Maria (2024). 'Greenland'. In Are Vegard Haug (ed.): *Crisis Management, Governance and COVID-19 – Pandemic Policy and Local Government in the Nordic Countries*. Cheltenham and Northampton, MA: Edward Elgar Publishing, pp. 105–112.

*Aftale mellem den danske regering og Naalakkersuisut vedrørende de særlige udenrigs -, forsvars- og sikkerhedspolitiske forhold, der knytter sig til udvinding og eksport af uran og andre radioaktive stoffer i Grønland* (2015). Grønlandsudvalget 2015–2016. Available at: https://www.ft.dk/samling/20151/almdel/GRU/bilag/33/1595416.pdf (Visited 4 October 2024).

Agreement on Cooperation on Aeronautical and Maritime Search and Rescue in the Arctic. Available at: https://oaarchive.arctic-council.org/items/9c343a3f-cc4b-4e75-bfd3-4b318137f8a2 (Visited 4 October 2024).

Arctic Hub. Available at: https://arctichub.gl/

*Bekendtgørelse af aftale af 31. august 2005 med Island om etablering af et fælles økonomisk område.* Available at: https://www.retsinformation.dk/eli/ltc/2013/13 (Visited 14 September 2024).

*Greenland in the World - Nothing about us without us. Greenland's Foreign, Security and Defense Policy 2024–2033*. Naalakkersuisut/Government of Greenland. Department of Statehood and Independence. Available at: https://paartoq.gl/wp-content/uploads/2024/03/Greenlands_Foreign_-Security_and_Defense_Policy_2024_2033.pdf

KNR-nyheder. Available at: https://knr.gl/da/nyheder/konflikt-i-nordisk-raad-faar-groenland-til-saette-samarbejde-paa-pause

Lackenbauer, Whitney P. and Rasmus Leander Nielsen (2022). '"Close, Like-Minded Partners Committed to Democratic Principles": Settling the Hans Island/Tartupaluk Territorial Dispute', *Arctic Yearbook 2022*, Briefing Notes. Available at: https://arcticyearbook.com/arctic-yearbook/2022/2022-briefing-notes/442-close-like-minded-partners-committed-to-democratic-principles-settling-the-hans-island-tartupaluk-territorial-dispute

Lov nr. 473 af 12/06/2009. Available at: https://www.retsinformation.dk/eli/lta/2009/473

Nordic Co-operation. News: https://www.norden.org/en/news/session-nordic-council-update-helsinki-treaty

*Udenrigspolitisk redegørelse 2003*. FM 2003/12. Nuuk: Landsstyret/Government of Greenland.

*Udenrigspolitisk redegørelse 2004*. FM 2004/20. Nuuk: Landsstyret/Government of Greenland.

*Udenrigspolitisk redegørelse 2005*. FM 2005/29. Nuuk: Landsstyret/Government of Greenland.

*Udenrigspolitisk redegørelse 2006*. FM 2006/18. Nuuk: Landsstyret/Government of Greenland.

*Udenrigspolitisk redegørelse 2007*. FM 2007/35. Nuuk: Landsstyret/Government of Greenland.

*Udenrigspolitisk redegørelse 2008*. FM 2008/41. Nuuk: Landsstyret/Government of Greenland.

*Udenrigspolitisk redegørelse 2009*. FM 2009/14. Nuuk: Naalakkersuisut/Government of Greenland.
*Udenrigspolitisk redegørelse 2010*. FM 2010/14. Nuuk: Naalakkersuisut/Government of Greenland.
*Udenrigspolitisk redegørelse 2011*. FM 2011/14. Nuuk: Naalakkersuisut/Government of Greenland.
*Udenrigspolitisk redegørelse 2012*. FM 2012/14. Nuuk: Naalakkersuisut/Government of Greenland.
*Udenrigspolitisk redegørelse 2013*. EM 2013/14. Nuuk: Naalakkersuisut/Government of Greenland.
*Udenrigspolitisk redegørelse 2014*. EM 2014/14. Nuuk: Naalakkersuisut/Government of Greenland.
*Udenrigspolitisk redegørelse 2015*. EM 2015/14. Nuuk: Naalakkersuisut/Government of Greenland.
*Udenrigspolitisk redegørelse 2016*. EM 2016/14. Nuuk: Naalakkersuisut/Government of Greenland.
*Udenrigspolitisk redegørelse 2017*. EM 2017/14. Nuuk: Naalakkersuisut/Government of Greenland.
*Udenrigspolitisk redegørelse 2018*. EM 2018/14. Nuuk: Naalakkersuisut/Government of Greenland.
*Udenrigspolitisk redegørelse 2019*. EM 2019/14. Nuuk: Naalakkersuisut/Government of Greenland.
*Udenrigspolitisk redegørelse 2020*. EM 2020/14. Nuuk: Naalakkersuisut/Government of Greenland.
*Udenrigspolitisk redegørelse 2021*. EM 2021/14. Nuuk: Naalakkersuisut/Government of Greenland.
*Udenrigspolitisk redegørelse 2022*. EM 2022/14. Nuuk: Naalakkersuisut/Government of Greenland.
*Udenrigspolitisk redegørelse 2023*. Nuuk: Naalakkersuisut/Government of Greenland.
WHO. Application available at: https://apps.who.int/gb/ebwha/pdf_files/WHA74/A74_INF5-en.pdf

# 6 Analysis

In this section we will look back at the beginning and try to answer the research questions posed. *What kind of activities are Greenland undertaken in international affairs? What are the objectives? Can Greenland be an influential player within IR? How far is the self-government stretched or limited in relation to IR?* We have already answered the first questions during Chapters 4 and 5 with Greenland's involvement in national and international relations by using the relationships between Greenland-Denmark, Greenland-USA, Greenland-EU, and the relationships towards international organizations and institutions as examples, where Greenland is using its maneuvering space in various ways. However, in this section we will use empirical evidence to put them into the theoretical lenses of foreign policy analysis, paradiplomacy, shelter theory and two-level games. The first section will use foreign policy analysis in relation to the MFA: s and what the outcomes they have achieved in the form of legislation or other matters over the 20-year period (recall Table 2.1 on page 14). The next section of paradiplomacy will take Greenland as a nation into account and its role in maneuvering within the foreign- and security policy. The third section is focusing on the Government of Greenland and how it functions in relation to the great powers in the Arctic, but also in relation to Denmark proper by using shelter theory and two-level games.

## 6.1 Greenland and Foreign Policy Analysis

Foreign policy analysis will here focus on the MFA: s and their achievements during the 20 years. Various legislative instruments seem to have been in focus to enhance Greenland's room for maneuvering within the foreign- and security policy area. As we have seen, major legislative changes have been made in the field of international relations. In 2003, we receive the Itilleq Declaration as a bilateral agreement between the Kingdom of Denmark and Greenland, where the two partners should be treated as equal partners in relation to foreign- and security policy, which is linked to Greenland or the Arctic. This was followed by the Igaliku Agreement in 2004 between the USA, Denmark and Greenland in relation to the upgrade of former Thule Air Base (now Pituffik Space Base). A year after in 2005, we see a new Danish legislation which also includes the Faroe Islands, which is the so-called Authorization Act, which gives both autonomous regions

DOI: 10.4324/9781003387657-6

*Table 6.1* MFA and Outcomes through 20 Years

| *MFA* | *Year* | *Outcome* |
|---|---|---|
| Hans Enoksen | 2003 | Itilleq Declaration 2003 |
| Josef Motzfeldt | 2004–2007 | Igaliku Agreement 2004<br>Authorization Act 2005<br>MoU with the USA 2006 |
| Hans Enoksen | 2008 | Ilulissat Declaration 2008 |
| Kuupik Kleist | 2009–2013 | Self-Government Act 2009<br>Kingdom of Denmark's Arctic Strategy 2011 |
| Aleqa Hammond | 2014 | Representation in Washington D.C. is opening 2014<br>Loss of service contracts in relation to the USA 2014 |
| Vittus Qujaukitsoq | 2015–2016 | Agreement with Denmark regarding Uranium 2015<br>Submission to the CLCS in 2016 |
| Suka K. Frederiksen | 2017 | West Nordic Council becomes observer within the Arctic Council 2017 |
| Vivian Motzfeldt | 2018 | Representation in Reykjavik is opening 2018 |
| Ane Lone Bagger | 2019 | Defense Agreement with Denmark regarding Kangerlussuaq 2019 |
| Steen Lynge | 2020 | Agreement Package with the USA 2020<br>Representation in UNPFII from Greenland 2020 |
| Pele Broberg | 2021 | Representation in Beijing is opening 2021 |
| Vivian Motzfeldt | 2022– | Agreement between Canada and Greenland regarding Tartupaluk – Hans Island 2022<br>Representation in NATO from Greenland 2023 |

within the Kingdom possibilities to sign bilateral agreements and enter international organizations as own nations. Furthermore, some MoU's have been struck between Greenland and other countries. In the table below we summarize the years with the respective MFA in relation to the outcome. This links all the foreign policy papers and the major outcomes over the years.

These outcomes are of course not only the MFA: s work over the years, since they have only happened to be sitting at their post while these outcomes have been implemented. The work regarding agreements and opening representation offices abroad and sending representatives to international organizations has been underway for several years prior to the realization of the outcomes. The fruits of earlier negotiations and legislation seem to place Greenland as its own nation within foreign- and security policy now with some restrictions due to the Kingdom of Denmark and maybe more because of the inference of the USA.

Another major change, which can be traced back to the 2000s is the evaluation of the Home Rule system and reports made, which in the end brought Greenland Self-Government in 2009 after a positive outcome in a referendum in 2008. The major change was to acknowledge the Greenlandic people as a people with the right to self-determination as it is stated in international law and the division of economy due to the takeover of natural resources. There is a phase out mechanism regarding how the distribution of incomes from exploitation of natural resources should be divided between Denmark and Greenland and how the block grant in this way should be phased out in the future. The Self-Government brought also a new structure and organization within the government of Greenland, where directorates became departments or ministries. The Greenlandic words for parliament (Inatsisartut) and government (Naalakkersuisut) became also official as the Greenlandic language itself. The work towards building up a Greenlandic diplomatic corps and representation offices abroad also became part of the enhanced self-government. This has helped Greenland to be able to negotiate and become an active actor within Arctic and other international relations.

A feature that has also been seen is boycotting as a tool to get the voice heard within different contexts in international relations. The first example with Aleqa Hammond boycotting the Arctic Circle General Assembly in Kiruna, Sweden back in 2013 and more recent Múte B. Egede's boycott of the Nordic cooperation within the Nordic Council and the Nordic Council of Ministers. It is hard to tell if this is something that has become praxis or not from the perspective of the Greenlandic government, since we have only experienced it twice during the 20-year period, but it seems to have an effect. It gives Greenland attention on the international scene and some solutions are then made as we have seen regarding the Arctic Council, even though this might be seen as a compromise and where Denmark is showing some signs of goodwill in relation to Greenland and the Faroe Islands. While writing this book, it is hard to tell if we will see any solution to the membership issue in relation to the Nordic Council and Nordic Council of Ministers. However, the Nordic countries are taking the matter sincerely and therefore we might expect some sort of compromise in the end. The discussion is ongoing about amending the Helsinki Treaty, which regulates Nordic cooperation.

## 6.2 Greenland and Paradiplomacy

Greenland as a nation has gone through a lot of changes over the centuries. We can say that WWII changed the dynamics within the foreign- and security policy area, since the USA took over as the protector of Greenland during the war. The connection towards Copenhagen was disrupted and therefore new influences from the North American side became part of the daily life. The Greenlandic people could for the first time trade with a superpower and have direct connections to Washington D.C. Greenland has been underneath Danish rule in over 300 years, but the changes from a colonial past to today have enhanced Greenland as becoming more and more independent especially within internal affairs, but also within international relations. We have witnessed activities within the indigenous rights movement from the 1970s and onwards, which has also strengthened Greenland as an international actor within the ICC and the UN framework. Greenland has given its voice and been able to be part of establishing various international organizations and been part of developing influential conventions and declarations, for example UNDRIP and ILO 169.

Greenland is using paradiplomacy to navigate within IR in various forms. Diplomacy and negotiations are used as the main channels when dealing with other countries based on values of human rights, indigenous rights and peace. Greenland wishes that the Arctic will still be a peaceful, low-tension area also in the future. However, some great powers such as Russia and China are pushing the other Arctic states (including Greenland) to be aware of military threats and other challenges that might spillover from wars and conflicts going on in other parts of the world.

The build-up of representation offices abroad is another channel where Greenland can act on its own and therefore negotiate directly with the EU, the USA, Canada, Iceland, and China in relation to trade, fisheries and education as the main areas of interest. These soft power areas can be seen as a tool for also enhancing relationships in other strategic areas in the future. As already mentioned earlier in the book, the security dimension with the USA is very well organized because of historical reasons, but the areas around critical minerals are also of crucial importance and Greenland has been able to protect its natural resources so far with best practices on an international level as well as harsh legislation on a national level.

As Greenland is developing its own strategies and is becoming more confident to navigate with great powers and other foreign powers, the development of various policy areas might become more intensified. This can be related to green transition, where Greenland has great potential because of a huge asset of hydropower and water assets from the melting ice. We have already witnessed that water might be seen as a new potential product to export as well as sand from the areas where the inland ice is retreating because of climate change. Another aspect of climate change is that the agricultural sector might become enhanced. For now, it is only in the southern part of Greenland that you can cultivate crops in larger amounts, while other parts in Greenland are using smaller green houses for salad for instance.

The huge infrastructure projects of extended airports in Nuuk and Ilulissat and a totally new airport in Qaqortoq will also change dynamics, since the world will become closer with connections with many parts of the world. This will enhance the tourism industry and is also seen as a potential gateway for closer cooperation within trade and business with other countries.

Greenland has also taken an active stand in participating in relevant international organizations and institutions to be able to get its voice heard and to secure that Greenlandic interests are considered. This takes the form of both intergovernmental organizations, parliamentarian organizations and other forms of transnational cooperation. The most prominent organizations and institutions are the Arctic Council, West Nordic Council, international fisheries and mammal organizations, the relationship towards the EU, the cooperation within the UN framework and recently NATO.

## 6.3 Greenland and Shelter Theory and Two-Level Games

Greenland has during the history used shelter from the USA and Denmark as the main shelter countries, but we have also seen that Greenland has seeking shelter from the EU, the Nordic countries, the UN and NATO in various ways. The Arctic Council and the international fisheries organizations can also be seen as giving Greenland some shelter. If we recall the differences between political, economic and societal shelter, we can draw some conclusions about the different degrees of shelter that is provided by the various actors and in what ways this shelter is utilized. The political shelter is about diplomatic or security where the small state is compensating for its limited resources in those areas; the economic shelter is about direct economic assistance, having a common currency or access to a common market, while societal shelter links to diffusion of ideas and values from foreign powers or international organizations. Table 6.2 illustrates the various forms of shelter and the author's estimation of the degree of shelter that Greenland seeks from each actor.

We have only used a binominal estimation of high or low degrees of shelter in this context based on the material which has been investigated during Chapters 4 and 5. This can of course be criticized to be too simple and not sophisticated enough, but it gives the reader an idea of how we can perceive shelter theory. From the table we can see that Denmark and the EU seem to be the most important shelter actors in this sense on all parameters with a high degree of shelter. This is not a surprise since both Denmark and the EU give Greenland political, economic and societal shelter in several ways. Politically Greenland stands very close to Denmark and is using the EU in diplomatic sense to negotiate favorable agreements. Economically both Denmark and the EU are supporting Greenland with huge amounts of financial support (i.e. the block grant, the partnership agreement and fishery agreement). Societal shelter is based on having the same values based on human rights and peaceful mechanisms and so on. The Nordic cooperation, the Arctic Council and the UN are all giving Greenland politically as well as societal shelter. This is based on the long history of cooperation with these actors and therefore the same values and ideas have been adopted in a Greenlandic context. The USA and NATO are both giving Greenland political or security shelter, while the international fisheries organizations are given Greenland some political shelter to prevent overfishing and utilization of Greenlandic waters in a destructive way. In this sense, Greenland can also protect its fishery industry from external influences.

If we turn to the two-level games, it is about the processes that Greenland has undertaken in order to sign bilateral agreements and other international agreements. We have witnessed two-level games taking place between Greenland and

*Table 6.2* Shelter from Various Actors in IR

| *Country/Organization* | *Political Shelter* | *Economic Shelter* | *Societal Shelter* |
|---|---|---|---|
| Denmark | High degree | High degree | High degree |
| USA | High degree | Low degree | Low degree |
| EU | High degree | High degree | High degree |
| Nordic Cooperation | High degree | Low degree | High degree |
| Arctic Council | High degree | Low degree | High degree |
| UN | High degree | Low degree | High degree |
| NATO | High degree | Low degree | Low degree |
| International Fisheries Organizations | High degree | Low degree | Low degree |

Denmark, between Greenland and the EU, and between Greenland and the USA, but also with other countries and organizations. It has not always been straightforward with negotiating agreements and coming to conclusions, but over the years we have witnessed that Greenland can now negotiate and conclude agreements on its own since the enforcement of the Itilleq Agreement and the Authorization Act. Before 2003 it was harder to conclude agreements, since Denmark was the only country to do so also on behalf of Greenland. The two-level games have therefore been enhanced over the years, while several bilateral agreements have been struck with foreign powers. There is however a catch and that is that the overall sovereignty when it comes to foreign- and security policy is still lying in the hands of Denmark, but Greenland can sign agreements within areas which is underneath its own competencies, i.e. regulated by the Self-Government Act.

As Olsvig (2022) has shown in her article closer and more direct dialogue has taken place between Greenland and the USA over recent years and this dialogue has not always included Denmark. This shows that Greenland can act alone as its own nation also in relation to the world's most powerful state (see Olsvig, 2022). As Greenland develops into more independence or where independence might become an option in the future, Greenland will be prepared for negotiations on a high-level with foreign powers in issues also related to security and defense. Several other two-level games might then occur, and it is hard to speculate what these might bring, but for now we can say that Greenland has learned to navigate within foreign- and security policy on the outset of this study.

## 6.4 Summary of the Findings

The major findings of this study are that we see that the role Greenland has taken within foreign- and security policy is an active one and that several outcomes have been achieved either by Greenland itself as a "state-like" actor or because of cooperation with others. Greenland finds itself as a nation following closely what is happening elsewhere and acts accordingly. If we use a universal policy steering model, we can conclude that it looks as follows, see Figure 6.1.

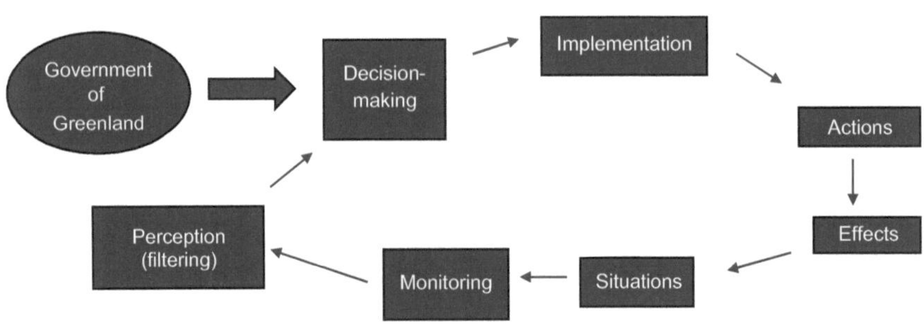


*Figure 6.1* Policy Steering Model for FPA.

*Source:* Jon C. Pevehouse and Joshua S. Goldstein (eds.). (2017). *International Relations*, 11th Edition. Pearson Publications, pp. 123–139.

The Government of Greenland makes decisions about its aims for foreign- and security policy, when these aims have been accepted by the parliament, the Government implement and begin with the actions towards reaching these goals. The effects then take place, and the consequences of Government actions can be seen in the public and in relation to other countries and international organizations. The situation is then monitored, and various perceptions might be visible through media and how other actors perceive Greenland's involvement and actions. It is a circulation of actions since the end goal will be to handle and adjust foreign- and security policy to new conditions throughout time. There will always be an overlap between political context, history and international relations within foreign policy analysis.

## References

Olsvig, Sara (2022). "Odd Couples' Win-Sets Maintaining U.S. Basing Rights through New Two-Level Game Negotiations with Greenland', *Scandinavian Military Studies*, 7:1, pp. 93–110. https://doi.org/10.31374/sjms.208

Pevehouse, Jon C. and Joshua S. Goldstein (eds.) (2017). *International Relations*, 11th Edition. London: Pearson Publications.

# 7 Outlook and Conclusions

We have seen a growing political, economic, and environmental interest in the Arctic region from a wide range of actors, which have been in focus in recent years. The Arctic region is experiencing some of the most rapid environmental, social, economic, and geopolitical changes, resulting from combined long-range pollution, climate change and growing tensions between great powers. This has affected Greenland and other states and nations in the Arctic. Greenland's geopolitical strategic position in the middle of the Arctic Ocean has therefore a role to play within international relations and it is therefore also a vulnerable position to uphold in relation to great powers. The study shows that Greenland has learned to navigate and maneuver within foreign- and security policy on its own and together with other actors. We have witnessed a tremendous development of Greenland as a subnational region with a high-level of self-government in areas that maybe not was seen possible a couple of centuries ago.

This study has been a journey into Greenland's international relations from various theoretical lenses and with different forms of documentation as sources. The study could have taken another turn with one specific theoretical lens as a main perspective to understand Greenland's role within IR to test this theory and by using other methods, but we have chosen to look at Greenland from various angles to give the reader a more holistic understanding of the dynamics that Greenland is experiencing within foreign- and security policy. The lesson learned from this is that Greenland can be seen as a case, which might be a forerunner for other subnational jurisdictions when it comes to how to deal with foreign- and security policy as a small nation in an ever more globalized world. To be able to decide if that is the case we would need to undertake a comparative study with similar entities to compare if this is true. This could be done for further research.

The study is limited to the framework that has been chosen by the author and therefore some critical points should be addressed. The idea to concentrate on only written sources is a limitation and could have been extended to include interviews, however several studies have already been undertaken interviews with the elite within the Greenlandic foreign- and security domain and since the same persons are rotating in the ministry and on high-level positions in the Greenlandic government and Inatsisartut it is hard to tell if this would have changed the picture of

DOI: 10.4324/9781003387657-7

Greenland as a case within this sphere. Another limitation is that the author is not having Greenlandic skills which is a shortcoming and therefore only Danish and English texts have been able to be interpreted. There might be nuances between the languages especially between Greenlandic and Danish, which therefore not have been able to account for. The book should prove to be useful in understanding Greenland and especially Greenland's policy within international relations and could therefore be used for teaching purposes whenever Arctic relations are discussed.

# Index

Note: **Bold** page numbers refer to tables; *italic* page numbers refer to figures and page numbers followed by "n" denote endnotes.